The Airwaves of Zion

The
Airwaves
of
Zion

*Radio and Religion
in Appalachia*

Howard Dorgan

THE UNIVERSITY OF TENNESSEE PRESS / KNOXVILLE

Library of Congress Cataloging in Publication Data

Dorgan, Howard.
    The airwaves of Zion: radio and religion in Appalachia / Howard
Dorgan. — 1st ed.
        p.   cm.
    Includes bibliographical references and index.
    ISBN 0-87049-796-0 (cloth: alk. paper)
    ISBN 0-87049-797-9 (pbk.: alk. paper)
    1. Religious broadcasting—Appalachian Region—Christianity.
2. Appalachian Region—Religious life and customs.   I. Title.
BV656.D67   1993
269′.26′0975—dc20   92-40704
                                                              CIP

To the faithful performers of
"The Morning Star Gospel Program,"
for forty-one years
of broadcasting
dedication

Contents

Illustrations

Illustrations

The Beginning

I enjoy telling audiences how I first encountered the "airwaves of Zion," a genre of religious broadcasting I define in the opening chapter of this volume. It was mid-August 1971. I had just moved my family to Boone, North Carolina, where I would be a professor in the Department of Communication at Appalachian State University. We had purchased a home, but could not take immediate occupancy; therefore, our realtor had arranged for us to spend two weeks in an A-frame summer house, resting on a ridge south of Boone, fronted with a deck that exposed us to a view of Grandfather Mountain, one of the most spectacular peaks in the Blue Ridge.

On the first Sunday morning of that stay, while the rest of my family still slept, I took a small radio out on that deck, leaned back in an unpainted slatted rocking chair, looked out across the softly rippled bluish haze that forged the line of a not-too-distant horizon, flipped to the radio's AM band where experience had taught me I would likely find touches of the region's local color, began my slow revolution of the tuning dial, and stopped when I heard the following: "It is now time for the Morning Star Gospel Program, sponsored and conducted by the Reverend Roscoe Greene. This program is on the air every Sunday morning at nine o'clock, with hymns sung by the Morning Star Trio."

I was listening to WATA-AM, Boone, North Carolina, and I was about to experience my first exposure to the locally produced live religious broadcasts that would engage my interest—off and on—for the next twenty years. In August 1991, those two decades later, "The Morning Star Gospel Program," would "sign off" for the final time, having broadcast for almost forty-one years to the residents of Watauga County, North Carolina. In addition, I would be in the WATA studios that August 1991, morning, concluding my fieldwork for this volume.

However, on the deck, that 1971 Sunday morning, I was receiving my initial introduction to diminutive and dedicated Sister Dollie Shirley (standing less than five feet tall but towering in her commitment to a personal mission), soft-voiced and moderately stoical Sister Hazel Greene (reading her announcements and prayer requests in a flattened voice that often concealed the emotions her tearing eyes would show), and impassioned and spiritually indefatigable Brother Roscoe Greene (the untutored Appalachian preacher who remains today as my prototype of the chanting exhorter on whom my descriptive studies were first focused), all airwaves-of-Zion personae who, from that moment would move in and out of my life as I became involved with the study of Appalachian religious traditions.

In one sense, then, my fieldwork for this volume started in August 1971. In a purer sense, however, it really got under way during the fall of 1988 when I began to identify the radio stations I would visit in my search for interesting and representative airwaves-of-Zion programming. Prior to that moment, my ethnographic studies had focused on a wider range of the Appalachian religious experience, starting first with my fascination with the region's preaching styles.

By 1988 I had already spent considerable time at three stations— WATA in Boone, WKSK in West Jefferson, North Carolina, and WMCT in Mountain City, Tennessee—those visits in pursuit of distinctive Appalachian sermonic styles exhibited by preachers in their live broadcasts. During the spring and summer of 1989—and later— I would extend this radio station field research into Virginia, West Virginia, and Kentucky, ultimately focusing on a relatively tight five-state area of central Appalachia. My goal became to find a limited number of Sunday-morning or -afternoon live religious broadcasts, originating from central Appalachian AM radio stations, with the respective airwaves-of-Zion personae and productions becoming subjects for this study.

By the spring of 1990 I had zeroed in on four stations, four weekly broadcasts, and four sets of performers. In all of these case studies, however, the actual airwaves-of-Zion program became only part of my total interest in the people involved. Each of the four studies pursues the respective personae outside of their airwaves-of-Zion environments, seeking to capture a more complete picture of the phenomena.

The reader may call this study "descriptive ethnography," or give it any other label he or she chooses. My purpose has been to capture, as vividly as I could, the Appalachian religious experiences I have witnessed, avoiding as much as possible both editorial judgment and academic theorization. That approach should not suggest any disdain on my part for the scholar who would examine the material from a more theory-based perspective. I simply pursue ethnographic ends most interesting to me and employ methodologies most suitable to those interests.

As has been the case in my previous Appalachian religious studies, I do not—except in response to specific requests or when sensitivities otherwise demand—preserve subject anonymity. My rationale is twofold. First, I want other scholars to be able to cover the same ground I have covered, should they wish to do so. Second, I am occasionally troubled by the scholarly aloofness—or the image thereof—that subject anonymity can suggest. In addition, I remember well a conversation I had with an Old Regular Baptist in Canada, Pike County, Kentucky, in which he told me that if I wrote about him he would appreciate the use of his real name. He wanted to be able to read specifically what I had to say about him. He was not belligerent, and it seemed a reasonable request.

Throughout this work I write in first person, occasionally reporting my own subjective reactions to people and events, while attempting to avoid the overtly judgmental approach mentioned above. Editorializing does come into play, however, since the very act of selecting a particular happening for description becomes a judgment in itself. The reader is exposed to the airwaves-of-Zion phenomenon through a channeling instituted by those selections and through the degree of emphasis I place upon any particular person, place, or spiritual performance.

When referring to the personae of this study, I employ the terms "Brother" and "Sister." It seems respectful to adopt titles they assign to each other within their own fellowships. My utilization of these terms, however, should not suggest any affiliation on my own part with the respective faith movement.

Recognitions

I need to recognize the assistance of some people, and I will start with individuals and organizations contributory to the funding of this project. In that regard I am indebted to Rolf Kaltenborn, trustee of the Kaltenborn Foundation, Palm Beach, Florida, who generously assisted in the financing of my research travels. I am also grateful to Loyal Jones and the Appalachian Center of Berea College, Berea, Kentucky, for an Appalachian Studies Fellowship Grant, also supportive of my fieldwork. Finally, I express my appreciation for the financial assistance I received from the Appalachian State University Research Committee, Linda Blanton, chair.

In this last regard, I want to recognize, with appreciation, the support of Joyce Lawrence, dean of the Cratis D. Williams Graduate School, Appalachian State University. For the last ten years, Dean Lawrence and her staff have been especially helpful in a number of my research efforts. Helpful, too, have been Harvey Durham, provost and vice-chancellor for academic affairs, and Ming Land, dean of the College of Fine and Applied Arts, both responsible for a sabbatical I was provided for the fall semester of 1990.

During the last three years I have visited an array of central Appalachian radio stations, and the station personnel with whom I have had the most contact have been the Sunday-morning and/or -afternoon announcers—frequently high school students, but at other times older persons with no formal training in radio—who provided information, taped broadcasts, introduced me to program performers, supplied station promotional materials, and performed a host of other services that aided this study. I am especially indebted to the following: Mary Lou Hayworth, WMCT, Mountain City, Tennessee; Wiley Vanover and Teddy Kiser, WNKY, Neon, Kentucky;

Barbara Justus and Alice Stiltner, WNRG, Grundy, Virginia; David Taylor and Ralph Wallace, WBPA, Elkhorn City, Kentucky; Martha Wilson and Jesse Corbett, WLRV, Lebanon, Virginia; Michelle Muncy, WAEY, Princeton, West Virginia; and Jeff Mullins, WDIC, Clinchco, Virginia.

My contact with station owners—except through phone calls— was limited to two individuals: Fran Atkinson, WMCT, Mountain City, Tennessee; and Sam Sidote, WELC, Welch, West Virginia. Atkinson initially had concerns about my work with her station, fearing I was about to ridicule her programming and all who worked for her. Her apprehensions were mollified after I talked to her about the attitudes behind my work. Sam Sidote, as will be shown in chapter 5, became a strong supporter of my study from the very beginning.

Obviously I owe a great deal to all the airwaves-of-Zion preachers, singers, musicians, and miscellaneous support personnel. With rare exceptions, these people were very open to my probings—and to my tape recordings and photography. I know I was often a decided distraction.

Ivan M. Tribe, University of Rio Grande, Rio Grande, Ohio, read an earlier version of my chapter on Rex and Eleanor Parker, providing not only a number of suggestions for that chapter but also a valuable collection of newspaper clippings. My colleague Danny Jacobson examined the material in chapter 1 concerning the changing nature of AM broadcasting, saving me from a couple of potentially embarrassing misstatements; and another colleague, Judy Geary, read several segments of this work, contributing valuable suggestions on form and style. A group of writers attending the 1991 annual meeting of the North Carolina Writers' Network provided an audience as I read several sections of the manuscript. I hope my reviewers make me feel half the elation I received from that audience's response.

Quentin Schultze, Calvin College, an authority on religious broadcasting, served as one of the readers for my submitted manuscript, providing particularly valuable advice relative to the conceptual base I outline in chapter 1; and an old friend, Deborah McCauley, Columbia University, also became one of the reader/critics for that initial manuscript. Add to the above all the good counsel I received from the staff of the University of Tennessee Press, especially Jennifer Siler, Meredith Morgan, Stan Ivester, and Kay Jursik.

An expression of appreciation to my family is always warranted. For twenty years, Kathy, my wife, and Shawn and Kelly, my children, have endured with grace the inconveniences my fieldwork has precipitated and the separations my withdrawals to writing engendered.

Finally, I offer words of thanks to three Appalachian religious figures who in the mid-1970s first got me interested in their particular airwaves-of-Zion broadcasts: Roscoe Greene, WATA, Boone; Dwight Adams, WMCT, Mountain City; and Earl Sexton, WKSK, West Jefferson. Each of these men has now devoted fifty to sixty years to his respective evangelistic mission, and collectively they have spent well over one hundred years in airwaves-of-Zion broadcasting.

—1—
The Airwaves of Zion
An Overview

> But ye are come to mount Zion, and to the city of the living God, the heavenly Jerusalem, to an innumerable company of angels,
>
> To the general assembly and church of the firstborn who are registered in heaven, to God the Judge of all, to the spirits of just men made perfect.
>
> Hebrews 12:22–23

Of all the metaphors employed within "plain-folk religion"[1] of the lower and upland South, there may be none richer than "Zion." In these southern regions—as well as in other parts of the nation—the word is used not to suggest a national Jewish homeland, or Jerusalem itself, but heaven and much more: a utopian afterlife and the longed-for, unimaginably wonderful "city of God" in which this afterlife is envisioned to be spent; an on-this-earth spiritual community of followers; the spiritual movement itself and all its component parts; an all-encompassing attitude, belief, and value system; and the intercommunion that connects all these elements—past, present, and future.

Therefore, throughout the South, and particularly in Appalachia, phrases such as the following are employed with great regularity: "children of Zion" (individual believers), "the church of Zion" (the collective body of followers), "songs of Zion" (hymns, particularly the older ones that are rich with tradition), "words of Zion" (either Scripture or preaching, but preaching that is more exhortative than instructional), "waters of Zion" (baptismal streams), "city of Zion" (heaven), and the phrase under present consideration, "airwaves of Zion."

Early settlers employed the rich metaphor "Zion" when naming their communities. There are a Zion, Arkansas, and a Zion, South

Carolina. There are a Zion Hill, Kentucky, and a Zion Hill, Mississippi. Then there are Zionville, North Carolina, and Zion Crossroads, Virginia. "Zion" also appears in literally hundreds of church names, of which the following are only a few from present-day Appalachia: Mount Zion, Old Mount Zion, Little Zion, Zion Ridge, Zion Hill, Zion View, Zion Grove, Zion Fork, Zion Meadow, Zion's Rest, Zion's Home, Zion's Way, and Zion's Hope, all of which are highly traditional "old-time" Baptist churches.[2]

In one way or another, "Zion" has come to symbolize the ultimate community of believers, the essence of God's residing place, the promise of eternity, the purity of traditional beliefs, and the beauty and blessings of faith. It is the verbal icon that cuts across all other facets of "plain folk" evangelistic Protestantism, integrating and unifying, legitimizing and sanctifying. The image thus projected is synoptic and much larger than, but inclusive of, the vision depicted in the well-known final verse of "We're Marching to Zion," a creation of the early eighteenth-century English theologian and hymnodist Isaac Watts, and a hymn that has become the clarion call to the endless and ageless "glory bound" march of the regiments of traditional Protestantism:

> Then let our songs abound,
>   And every tear be dry;
> We're marching thro' Immanuel's ground,
>   We're marching thro' Immanuel's ground,
> To fairer worlds on high,
>   To fairer worlds on high.
> We're marching to Zion,
>   Beautiful, beautiful Zion;
> We're marching upward to Zion,
>   The beautiful city of God.[3]

"Airwaves of Zion"
A Definition

I begin this opening chapter with an effort to define the phenomenon under study. Defining is often an artificial and arbitrary process at best, but one that becomes necessary when we communicate with imprecise phraseology, in this instance the term "airwaves of Zion."

The phrasing is not mine. I wish it were. Unfortunate for that wish, however, is that I have heard it perhaps a hundred or more times during my two decades of scrutinizing Sunday religious broadcasts from radio stations in central Appalachia. Furthermore, I have heard the expression elsewhere in the South. In some circles the phrase is commonly understood and commonly employed.

For me, the terminology is much richer, more colorful, and even more exact than others I considered using in this work: "Radioland Religion," "The Appalachian Electronic Church," "The Appalachian Church of the Air," "Appalachian Airwaves," "Mountain Radioland," and "Mountain Airwaves." In addition, I consider the phrase more respectful than the ones above, coming as it does from the rhetoric of the phenomenon itself.

Airwaves of Zion—*as I am limiting the phrase for the purpose of this present study*—relates to a genre of locally produced live religious broadcasting that emanates from the AM stations of Appalachia; on Sundays these stations air a string of programs of preaching, singing, testifying, praising/glorifying, and other types of religious expression, all colored with a heavily provincial, fundamentalist, usually millenarian, "Come to Jesus" evangelism. It is a phenomenon relating primarily to rural areas, but not necessarily so; it is a division of religious broadcasting that draws its supporters (performers and listeners) principally from older and/or less-well-educated individuals who hold to religious beliefs and practices generally identified with a lower- to lower-middle-class evangelical Protestantism; and it is an electronic church segment that lies snugly within America's "folk religion" base.

Defining Characteristics

To summarize and extend, I now put forth the following twelve statements as defining characteristics of airwaves-of-Zion programming:

(1) It is not mainline in its denominational ties but is produced by individuals and groups from such denominations and sects as Holiness-Pentecostal (most of them independent and representative of the interesting blending—in Appalachia—of these two theological movements); Church of God (the Cleveland, Tennessee, branch); Assemblies of God (rural units thereof); Freewill (or Free Will) Baptists (again most of the groups being nonaligned, at least with any established organization such as the John-Thomas Association of Freewill Baptists); Independent Baptists; and a host of nondenominational, wholly autonomous churches that might identify themselves by such titles as "Full Gospel" (usually signifying Holiness-Pentecostal lean-

ings), "Church of Signs and Wonders" (the latter terms again suggesting Pentecostal influences), or "Church of Prophecy."

(2) It is not a product of larger institutionalized structures such as conferences, associations, synods, dioceses, and the like, but draws its impetus from the very personal motivations of individuals and groups—who may represent a church, but who more often simply represent themselves, assuming all of the responsibilities of the program, financial and otherwise, and exercising full control over the theological messages.

(3) It is produced by preachers and singers who have had no formal training in broadcasting and who possess varying, but generally deficient, degrees of electronic communication expertise, finesse, or knowledge of mass media law and ethics, factors occasionally embarrassing to their respective stations.

(4) It is noted for its improvisational tone and its "led by the Spirit" spur-of-the-moment happenings, encouraging full and free participation not only within the performing unit but also from listeners, through phone-in requests or responses and by encouraging them to drop by the studio.

(5) It is broadcast over pay-for-time commercial stations and is—with some exceptions—financed by "freewill offerings" and "gifts of love."

(6) It is strongly personal in its themes and messages—testimonial and confessional, naming sins and sometimes sinners, often addressing specific communities of listeners, and frequently responsive to such "real-life" travails as losing one's health and/or financial security, finding one's self left alone in life after a spouse and other kin have "passed over yonder," being concerned about the physical or spiritual well-being of loved ones, and experiencing doubt concerning one's own spiritual condition.

(7) It is highly emotional and cathartic, producing in-studio scenes of shouting, weeping, hugging, and arm waving, or other unbridled kinetic expressions.

(8) It is as serviceable to the needs of the studio participants as it is to the needs of the airwaves-of-Zion listeners, if not more so.

(9) It has no connection with, nor is it comparable to, either that more sophisticated, high-tech and high-financed phenomenon we call "televangelism," or that religious-programming-only radio movement that has produced numerous "all-Christian" stations throughout the nation.

Instead, (10) it falls clearly within the genre of spiritual expression

Harrell calls "plain-folk religion" (mentioned at the beginning of this chapter) and—in a more qualified way—under the broader heading of "folk religion."

(11) It is especially rich in Appalachia, where some pockets of traditionalism have outlived comparable phenomena in other regions of the nation.

Nevertheless, (12) it is now a dying phenomenon, falling prey to changes in both the broadcast industry and the general cultural environment in which it has been found.

Defining Characteristics Explained

The following discussions examine these defining characteristics in more detail. In addition, examples from the phenomenon are provided where needed for greater clarity.

Not a mainline movement. The airwaves-of-Zion phenomenon is not something supported by such large denominations as the Southern Baptists, the United Methodists, the Presbyterians, the Lutherans, the Episcopalians, and the like; nor is it tied to or produced by other organized religious units such as the associations to which "old-time" Baptist groups (Regulars, Separates, Primitives, Uniteds, etc.) traditionally belong. In addition to emerging from the various "plain folk" evangelical groups, this broadcasting genre—in most instances—is fostered primarily by individuals and groups who practice their callings independent from any institution larger than a single church. Indeed the preacher involved is often operating from a totally nonaligned and free-lance position, supporting the respective broadcast through his or her own incentive and out of his or her own pocket, trying to generate enough money through those "freewill offerings" to meet the cost of airtime, or demonstrating a willingness simply to pay to be heard.

Not a product of larger institutionalized structures. This absence of any tie to mainline denominationalism parallels a strong trend in Appalachia toward independence for the local church, a factor that is particularly true among the Holiness-Pentecostal elements of this phenomenon. The term "Holiness-Pentecostal" is employed in this volume to stand for a wide range of highly individualistic, nonaligned fellowships found throughout southern and central Appalachia; these fellowships have combined elements of the Holiness and Pentecostal faiths to create a mixed and greatly varied theological base. In this work I hyphenate the term to suggest this amalgamation of what started as two separate religious movements. The term is not employed here to signify a formal denomination.[4]

This strong tradition of local-church autonomy is also true for the Independent-Freewill-Missionary Baptist elements of the movement. In general, Appalachians have demonstrated considerable reluctance to surrender the freedoms of their local churches to the hierarchical structures of conferences and associations. Indeed, this reluctance precipitates frequent battles within Appalachian "old time" Baptist groups for whom associations exist as part of the institutional structure.

The lack of any tie to larger institutions of ecclesiastical governance often results in instability in the phenomenon. One manifestation of this instability is that these small AM stations often experience a high turnover in their Sunday programming. A preacher may organize a group of musicians and singers, establish a broadcast, keep it going for six months or a year, and then find the task to be more taxing (financially or otherwise) than he/she originally projected. When the respective program is withdrawn, a "transitional station" (examined later in this chapter) might decide not to fill the time slot with another airwaves-of-Zion group, opting instead for thirty more minutes of gospel music. In the case of those AM stations still firmly fixed within this airwaves-of-Zion phenomenon, another preacher and singers may be standing in line for the next available opening.

Some programs, of course, have defied this tendency toward temporariness, staying on the air for many years and surviving a number of generational shifts in personnel. "The Morning Star Gospel Program," examined at the close of this chapter, was on WATA, Boone, North Carolina, for forty-one years. At WMCT, Mountain City, Tennessee, Brother Dwight Adams kept a program going over that station for twenty-two years before he retired from the active ministry. He had earlier devoted ten years to the same program, but on another station. Brother Douglas Shaw is currently in his nineteenth year of broadcasting at WMCT. Finally, Brother James Kelly, WNKY, Neon, Kentucky, has been at that station for almost forty years. He has not been able to remember precisely when he began.

The absence of a tie to any established organizational unit (larger than that single church) also results in a great diversity in theological positions, as individual preachers interpret the Scriptures and traditional Christian doctrine to meet their own experiences and needs, the experiences and needs of their cadre of helpers, and the experiences and needs of the particular "radioland" listeners. Because the groups are evangelistic, however, the homiletic rhetoric stays to the Arminian/general-election side of the atonement-doctrine continuum, thus proclaiming that Christ died for all men and not just for a chosen few, the elect. Those more Calvinistic/limited-atonement preachers (Primitive Baptists, Regular Baptists, Old Regular Baptists, and the like) generally avoid taking to the air with their messages for fear of violating the "God calls, not man" principle.[5]

In addition, there is a heavy Pentecostal influence within the theological base of the airwaves of Zion, even among the Freewill, Missionary, and Independent Baptist groups, with the term "full gospel" often becoming a pivotal phrase. In chapter 2 of this work, for example, we see this term become critical to the beliefs of Brother Johnny Ward. Under the phrase he houses such basic tenets as (1) the inerrancy of Scripture, both New and Old Testament; (2) the immutable sinfulness of man, apart from God; (3) the doctrine of salvation and justification by grace; (4) the outpouring of a "Latter Rain," or second Pentecost; (5) the premillennial "second coming"; (6) the possession by the Holy Spirit and the special blessings ("powers") that subsequently ensue—divine healing, Spirit-possessed witnessing, "tongues," and other "signs and wonders"; (7) Hell and Satan—the "lake of fire"; and (8) heaven, eternal life, and union of the Saints in Zion.[6] Furthermore, in each of the other three case studies readers will see some Pentecostal influence.

No formal training in broadcasting. I have noted elsewhere the reluctance of the more traditional Appalachian preachers to subject themselves to any kind of theological or homiletic schooling, believing that such education would not only be useless but threatening to their "old-time" ways—useless in light of their belief that when God calls he also equips, and threatening in the sense that they view formal theological training as the fostering of a multitude of extra-gospel doctrines and practices.[7]

Appalachian radio preachers also see no value in formal broadcast training, having seen scores of their fellow exhorters, inspired by the Spirit, develop airwaves-of-Zion programs without such advantages. Furthermore, the station managers involved with this phenomenon make no training demands upon the preachers, usually satisfying the stations' concerns through the pronouncement of a few basic station policies similar to the following hypothetical ones: Do not get on the air and slander other religious groups; do not involve the program, and thus the station, in any sort of racial, ethnic, or nationality attacks; and do not wage any campaigns against any other individuals or groups in the community that may be embarrassing to the station. Even this minimal level of instruction, however, it not always provided, and I have talked to airwaves-of-Zion preachers who claim to have been given absolutely no information relative to station policy or federal broadcasting regulations.

It needs to be said, however, that this genre of Appalachian religious broadcasting is not particularly noted for vitriolic attacks on other factions, particularly other religious groups. Some negativism in found in the rhetoric of the phenomenon, but it is aimed more at ideas and practices than at individuals. For example, these preachers will show no fondness for alcohol sales, lottery tickets, roadhouses they consider profligate, magazines they judge to be pornographic, movie and television entertainments they view as improper, abortion, lack of prayer in public schools, and a host of similar social and political factors; but they will seldom direct their attacks at the person or persons involved rather than the ideas, products, or practices. In other words, these exhorters typically do not belong to the school of religious attack-broadcasting exemplified by Father Charles Coughlin, Reverend Gerald L. K. Smith, and Reverend "Fighting Bob" Schuler.[8]

Part of the reason for this general avoidance of attack-broadcasting probably can be attributed to the Appalachian live-and-let-live attitude, but the rest can be credited to the "come to Jesus" evangelical mission of these preachers. "My aim is to preach Jesus and his love," said Brother Dean Fields of Letcher County, Kentucky, "not to divide folks. I want my Brothers and Sisters to love me and everything I stand for." "I work with all preachers who will work with me," proclaims Sister Brenda Blankenship, a Pentecostal evangelist from Premier, West Virginia. "God's love is a pretty big umbrella."[9]

The various religious broadcasting techniques practiced by these airwaves-of-Zion preachers—regarding preaching, gaining listener involvement, and generating contributions—are learned through experience and informal apprenticeships. Like their homiletic skills in general, the studio methods these preachers employ are developed by hearing and watching others, and occasionally by starting off as a helper on some other program. It is not surprising, therefore, that there is much similarity between program formats, the radio preaching techniques, and the methods of audience engagement.

Indeed, program formats are often almost mirror images of each other, involving as they do such common show segments as an opening theme hymn; a beginning prayer; more hymn singing; the urging of call-in requests or responses; announcements of regular church meetings, special services, and other religious happenings; acknowledgments of studio visitors, infirm broadcast personnel or supporters, contributors, regular listeners, or just friends and relatives of the program's principals; preaching; and brief appeals for support, financial or otherwise, before going off the air in song and/or prayer. The four case studies developed in this volume do involve airwaves-of-Zion programs that possess their own distinct formats and styles, but that distinctiveness does not belie the considerable similarity within the broader phenomenon.

One airwaves-of-Zion program practice, on-the-air acknowledgments of scores of people, becomes an especially important part of these formats. This procedure involves the recognition of individuals and groups who have sent in contributions, written those letters of support, or simply become regular listeners. Such persons evidently enjoy hearing their names announced over radio, and some preachers spend sizable segments of their airtime calling out long lists of people to be "remembered in prayer," thanked, or "blessed," building or preserving listener audiences in the process.

This was the case when I visited Brother Rick Thompson's "Light of Life" program, broadcast each Sunday over WLRV in Lebanon, Virginia, and when I observed an airing of "The Gospel Light Broadcast," WAEY, Princeton, West Virginia, led by a preacher whom I am able to identify only as Brother Hall. Both Hall and Thompson devoted extensive portions of their programs to a calling out of names, with added comments about each person's condition or relation to the program. Hall, however, included more names on his broadcast than any Appalachian radio preacher I have observed, introducing well over one hundred during his program.

First of all, Brother Hall's "Gospel Light Broadcast" is an hour-long show, and he spends a major portion of that airtime keeping his listeners aware of circumstances—primarily of a health-related nature—in the lives of his friends, acquaintances, and program supporters. He does this with the help of several pages of longhand notes that he spends much of his week compiling as he visits various hospitals and nursing homes in the region—who's in what hospital, under what circumstances, and encountering what degree of progress in her or his illness. Brother Hall is an elderly and noticeably empathic pastor-type preacher, and all of this information is related to the audience in a voice that is kind and comforting, but decidedly slow, low-key, and lacking dynamism. The result is a droning-on effect that must have little interest to individuals who do not belong to Hall's "radioland" community of followers.

As an illustration not only of this preacher's heavy emphasis on people and their circumstances but also of the general tone of his announcements, I will record here the first few moments of his program as it developed on April 16, 1989, one of the four times I visited the WAEY broadcast facilities. During this segment of his "remembering" people, he included individuals from only one health-care facility in the region. Later he covered other such units, following the basic pattern he established in this opening period. Readers should imagine a soft-spoken, slow, and sensitive vocal style, with slightly increased emphasis placed upon the first pronouncement of each name, the speaker making certain that his audience hears the respective name.

A very pleasant good afternoon to all our friends and neighbors out along the way. It's always a joy to come your way each Sunday from one to two o'clock with the "Gospel Light Broadcast," along with our special singers, Brother and Sister Wiley.

I want to say now that the first thirty minutes of our broadcast is being dedicated to Mr. and Mrs. Harley Wyatt, McNutt Avenue, here in Princeton, good friends of ours. We are also remembering Mrs. Elvita Thomas in the first thirty minutes of the broadcast, and several of our friends that's out in the Princeton Community Hospital we'd like you to remember in prayer.

Mrs. Flossie Crouch has been there now a few days. She underwent surgery. And I was in—I believe it was Friday evening—to see her, and she's getting along some better.

Now our good friend Warner Burkett is still in the Princeton Community Hospital. Mrs. Julie Evans had surgery one day this week. Seems to be getting along good. She's in the ICU. And Mrs. Nora Tiller, a aged mother from Lashschmit [spelling?] is also a patient. And Mrs. Francis Lizzie Clemens, from down I believe on Piggett Creek, off Oakdale, near that vicinity. And this aged mother has been a very sick lady.

And Brother Garfield Stewart's wife, she's been there now for several days. And this lady's been down for quite a while. And we do ask you to remember Brother Garfield Stewart's wife and also remember Brother Garfield Stewart. He's an awful good man, a good friend of mine.

Mrs. Louise Carico—remembering her and also her husband. Mrs. Carico has had surgery, and her husband had a bypass down at Duke University Hospital.

Then our good friend Walter Elmore, our neighbor for quite a while. And he had—I believe was—a heart attack, along this past Friday.

Then, Mr. Meadows, also a patient there. And Mary Comer is in Princeton Community Hospital.

And I want to remember a lady that's been discharged from the hospital, Mrs. Eula White.

And let me announce now a revival meeting which I'll begin on tomorrow night at the Christian Mission, located on the road between Littlesburg and Brushfork, service each night at 7:00 P.M. We're expecting special singing each night. Brother Wade Bishop is the pastor, and I trust you'll be in much prayer for the revival meeting. We'll see people saved, the church will be blessed, and we can all then rejoice together in the Lord.[10]

Another way in which names of listeners are introduced is through dedications. A hymn will be announced, and then several names will be read as persons for whom the selection will be dedicated, occasionally with comments that explain the reason for the dedication. In the early 1970s, when I first began following WATA's "The Morning Star Gospel Program," I would often be amused when I would hear Sister Hazel Greene, wife of the preacher, Roscoe Greene, announce a hymn with a statement similar to the following: "The title of our next hymn is 'Amazing Grace,' and it goes out to Brother and Sister _____ who last week was in a car accident up near Beaver Creek in Ashe County. They wasn't hurt, and we praise God for that, but I understand they don't have no car now. So the folks in their church might think about that and swing by their way on the way to services." Then as soon as Sister Hazel would finish, Sister Dollie Shirley, who led all the singing for that show, would add one or more dedications, complete with intimate details about the people she wanted listeners to "remember." Next the other two members of the Morning Star Trio might throw out additional names and narratives. The result would be that by the time the hymn was sung it was "going out" to seven or eight people, about whom the audience had learned some current information, of either a sad or happy nature.

I said I was often amused by such program episodes. Later, however, that amusement was changed to a deep respect for the tradition, recognizing the role such reporting played in the lives of the listeners. Acknowledgments and dedications constitute, of course, a significant part of that broadcaster/listener intimacy that prevails throughout the airwaves-of-Zion environment, and it is this intimacy—this tradition of strong listener/broadcaster connectedness—that stands most in danger as small-town radio goes through the transitions of the moment.

Improvisational in nature. One characteristic of this genre of religious broadcasting that most intrigued me when I first began my observations was the relaxed, improvisational nature of these programs. In *Giving Glory to God in Appalachia* I mentioned the seemingly unstructured nature of many airwaves-of-Zion shows, a prevailing characteristic that occasionally permitted people to enter the studio late in the broadcast and immediately be asked to step to the mike for a testimonial.[11] This lack of structure is also seen in the selection of songs and singers, individuals who will lead prayers, those who will testify, the recipients of dedications and remembrances, and the person or persons who will preach. No format is so set that it cannot accommodate "Spirit-led" developments.

There is always one rationale behind all this improvisation, the conviction that anything too tightly planned by man precludes the involvement of God. So these airwaves-of-Zion performers inevitably leave some openings for the "Spirit" to get through and redirect, thus confirming a belief in the old adage that "man proposes and God disposes." "I always know that something will fill the broadcast," Brother Roscoe Greene once said, "because God doesn't let us down."[12]

One element that helps in the filling of these broadcasts is of course the singing, and here improvisationalism is certainly at work as last-minute decisions are made concerning what hymns will be sung and who will sing them, decisions that in turn are reversed five minutes later when a particular circumstance arises or when a particular singer enters the studio. This is certainly so in the "Words of Love" broadcast, being examined in chapter 4. Brother Dean Fields, the moderator of that program, usually has a dozen or more singers standing ready to perform, but he will always make room for a person who drops by after being away from the studio for several weeks. His motive, in part, seems to be to bring the person back into the fold, to let that singer know that he or she has been missed.

The same principle often seems to apply to the question of who will preach the ten or so minutes devoted to exhortation. In the late 1970s, when I regularly followed "The Morning Star Gospel Program," Brother Roscoe Greene had five or six preachers who occasionally dropped in on the broadcast. These were men who pastored a collection of small churches whose members were particularly supportive of the program. When any one of these individuals would visit the show he would routinely be invited to preach. Indeed, the prevailing protocol seemed to mandate such an invitation.

There are also some broadcasts that are jointly sponsored by two or more preachers, any one of whom might speak on a given Sunday. This appeared to be the situation for "The First Freewill Gospel Time," WMCT, Mountain City, Tennessee; "Keep on the Praying Ground," WNKY, Neon, Kentucky; "The Mountain Empire Full Gospel Hour," WDIC, Clinchco, Virginia; "Heart to Heart," WELC, Welch, West Virginia; and "Hollis Ratliff and the Victory Gospel Singers," WBPA, Elkhorn City, Kentucky. In the last two examples, two or more women regularly stand ready to preach.

Airwaves-of-Zion preaching itself is always improvisational. I have never seen exhorters on any of these programs use notes or a manuscript, and there is probably no thesis more entrenched in the concords of this movement than the one proclaiming improvisation as essential to truly "anointed" exhortation. "Spirit" speaks, so declares the principle, only when "nature" steps aside and permits the "revelation." "I'm just gonna stand here and let the Lord have his way," proclaimed Brother Garrett Mullins during his "Jesus Is the Way Broadcast" over WNRG, Grundy, Virginia. "When he shuts up, I shut up."[13]

Such statements should not be interpreted to mean these preachers believe all improvisational homiletics result in God-proclaimed rhetoric. These preachers acknowledge the existence of some exhortation that is "blessed" ("anointed," "revelated," "of the Spirit") and some that is not. Their problem, of course, is to tell the difference. An indefinable, yet recognizable, quality in the preacher's passion—real, not just bombastic—tells the story, they say. According to this premise, "Spirit" cannot be counterfeited. Such a statement obviously becomes problematic, however, when two preaching Brothers disagree on the "blessed" state of a third exhorter.

Listener responses also play a role in this freewheeling, play-it-by-ear, improvisational program atmosphere, with listeners' calls occasionally changing the direction of a program. Preachers like Brother Johnny Ward of WMCT, Mountain City, Tennessee, and Brother Dean Fields, WNKY, Neon, Kentucky, make frequent appeals for audience responses in the form of prayer requests and the like, and listeners do call the station with their particular appeals or comments. Usually these auditor contributions are relayed to the preacher by the person on the main board (through written notes), but in the case of Fields's show a church member tends the phone throughout the program, occasionally calling Fields himself to the phone at times when singers are performing. As we will see in chapter 4, such calls may precipitate shifts in program direction.

Such listener responses contribute heavily to that previously mentioned broadcaster/audience intimacy that characterizes airwaves-of-Zion programming and airwaves-of-Zion stations. When I was visiting WBPA in Elkhorn City, Kentucky, I spent some time, before the live programming began, chatting with David Taylor, the Sunday-morning announcer and gospel-music disc jockey. Taylor was in the process of airing gospel recordings as requests came in for the particular releases. He had received a request for and played a comic gospel piece that I think was titled "I Ain't Gonna Handle No Snakes," a spoof about a man who accidentally found his way into a serpent-handling service, a highly exotic religious practice found in limited areas of Appalachia.

No sooner had Taylor removed the record from the turntable than a car pulled up in front of the station. A woman got out of the car, marched into the station and into the main studio, informed Taylor that she was Holiness-Pentecostal, and let him know, with very direct language, that she found the recording disrespectful of her beliefs and worship practices.

Here, of course, was a situation in which a station announcer found himself caught between listener responses—one requesting the recording and another denouncing it. I asked Taylor if he intended to pull the release from the collection of singles and albums regularly played, and he said he would need to check with the station's program manager before taking that action. That's where my information relative to the event ends, but the scene has remained in my mind as one of the most graphic examples of listener involvement I have observed.[14]

Broadcast over pay-for-time stations and financed by freewill offerings. Currently there is no incentive for stations to provide free time for this airwaves-of-Zion programming: as I will discuss later in this chapter, Federal Communication Commission (FCC) regulations relative to the reporting of percentages of time devoted to public service programming and announcements have been terminated. The result has been that these preachers and singers do have to pay for their airtime, whereas at one time at least some of this airtime was free. Nevertheless, at the airwaves-of-Zion stations I have visited, the amounts are small, generally ranging (at the time of this writing) from twenty-five to fifty dollars for a thirty-minute time slot. WNKY in Neon, Kentucky, charges Brother Dean Fields two hundred dollars a month for his use of four one-hour time segments, one hundred dollars of which is paid by one generous member of Fields's church. WMCT, Mountain City, Tennessee, bills its half-hour preachers thirty-five dollars a week, and at the time Brother Roscoe Greene closed out his tenure at WATA, Boone, North Carolina, that station was assessing the "Morning Star Gospel" group fifty dollars a week for their 9:00–9:30 Sunday-morning show.

I will discuss the money-raising techniques of these airwaves-of-Zion preachers under another heading, and for now I will only mention that few of these individuals become excessive in their appeals for financial support. Since the broadcast-time fees are relatively small, there is no need for the constant begging for money that has become so characteristic of much of televangelism. During the twenty years I followed "The Morning Star Gospel Program," I do not recall ever hearing Brother Roscoe Greene ask specifically for money. Instead, he would say something like the following: "Let us hear from you to know you listen to the program. It's always a blessing to receive your letters." Some preachers rely on nothing more than an announcer's opening or closing statement: "This program is supported by freewill offerings."

Strongly personal and direct. In *Giving Glory to God in Appalachia* I related a story about Brother Albert Tester of Sugar Grove, North Carolina, who, in November 1973, filled part of his airtime at WMCT, Mountain City, Tennessee, by talking personally to a Robert Morefield, whom Tester identified—over the air—as being "unsaved." During that broadcast Brother Tester went on, in this very public way, to counsel Brother Morefield concerning the state of his soul and what he needed to do "to be saved." I used the episode to support my contention that Appalachian religious rhetoric is usually very direct, always naming the sin and sometimes naming the sinner.[15]

Since that writing I have encountered numerous other examples of this tendency toward directness and intimacy, ranging from the simple act of identifying an individual said to be experiencing problems to a full detailing of those problems. The most innocuous of such public proclamations might be Brother Hall's detailings of listeners' medical circumstances (WAEY, Princeton, West Virginia), while Brother Dean Fields's identification of a named individual's alcohol dependency (WNKY, Neon, Kentucky) could be considered considerably more personal, and, by the standards of many people, improper.

Brother Fields, however, would never consider such a naming improper, as long as that individual was publicly striving to free himself or herself from the dependency.[16]

Readers need to understand the ideological and theological base from which Brother Dean Fields operates. First, he believes that all persons are innately sinful and that the only thing an individual can do is let "Jesus wash away" the old sins and help control the new ones. Second, an old sin abandoned is a mark of pride rather than a mark of shame; therefore, as we will see in chapter 4, Fields frequently proclaims his own earlier weakness for alcohol, on the way to celebrating his current freedom from that weakness.

In the course of my fieldwork for this volume I heard names and personal problems being aired over every station I visited, and in a number of cases the narratives that followed were about matters that could be considered extremely private. I heard the name of a woman whose husband had abandoned her, leaving her with two children; I found out that a particular Sister's son had been incarcerated, with some details of the crime; I learned that a certain Brother was struggling with the state of his soul; and I learned about two named parents who were distraught over the birth of a physically deformed child. Furthermore, given the understanding shared by these airwaves-of-Zion participants and the openness of these broadcasts, none of these revelations seemed especially inappropriate and offensive. Nevertheless, I believe it safe to assume that a vast majority of my readers would not like their names and personal difficulties revealed in such a public way.

This same degree of directness and intimacy generally prevails among the broadcast participants, with sermons, testimonials, on-the-air conversations, and announcements often telling audiences much about the lives of the studio personae. "We're family," says Brother Roscoe Greene, "and we have struggled together."[17]

I do not want to show disrespect for these broadcasts by comparing them to soap operas, but there is much about the nature of this airwaves-of-Zion programming that allows listeners to keep up with the life stories of a show's principals. Of course, this is especially true of a long-running broadcast like "The Morning Star Gospel Program" became. During the 1980s, for example, I followed the protracted illness and eventual death of Sister Dollie Shirley's husband, gleaning the narrative from weekly prayer requests.

Highly emotional and cathartic. Given this intimacy within the performing groups, these airwaves-of-Zion studios are frequently flooded with emotionality, as scenes of crying and rejoicing reach the highest possible levels of expression. In turn, these emotions generate deep catharses that seem to end in equally high levels of tranquilization.

At WNRG, Grundy, Virginia, Brother Garrett Mullins, the Pentecostal pastor of the Fountain of Life Church, also in Grundy, comes on the air (at the time of this writing) each Sunday afternoon at one o'clock. He is assisted by Brother Gleasen Mays, who plays an electric guitar, and three singers—Sisters Diane Mays, Cathy McGlothim, and Pam Fuller, whom he identifies as the Fountain of Life Church Choir. Mays brings with him an amplifier and two speakers, preparatory for filling the studio with a sound volume far in excess of what is necessary for broadcasting. As we will see in chapters 2, 3, 4, and 5, highly amplified preaching and singing seem essential to some of the worship styles present in and around the airwaves-of-Zion scene, especially the Pentecostal segments of that scene.

The WNRG studio that is devoted to airwaves-of-Zion programming is larger than most of the studios I visited during this study; consequently, Brother Mullins has considerable space in which to move while he is preaching. In addition, the main mike is attached to a cord that allows at least fifteen feet of play. All of this works well for Mullins's preaching style, since he is seldom in one spot for longer than a few seconds—sprinting, bounding, crouching, and leaping, all accompanied by exaggerated movements of his arms, shoulders, and head.

Typical of Pentecostal preachers, Mullins draws fervor from emotional displays around him, moving close to anyone who might respond dramatically to his exhortations. It is in this regard—perhaps more so than through their singing—that the three women contribute to the emotionality of this broadcast, crying, shouting, wailing, and occasionally laughing joyfully.

Sister Cathy McGlothim is particularly active in these responses, displaying emotions quickly, forcefully, and with every facet of her body—waving her arms above her head, clapping her hands together, performing toe jumps as she claps, and occasionally standing fixed in a particular state of excitation as Mullins loudly proclaims his message inches from her face. Indeed, it is during Sister McGlothim's relatively immobile moments that she nonverbally communicates with the greatest intensity.

During such episodes there is no indication that McGlothim receives this focus because she is the person most in need of an evangelistic message. Instead, every aspect of the situation suggests that Mullins experiences fervor in direct proportion to his audience's intensity of response. By this process, the small group of enthusiasts, and especially Mullins and McGlothim, generate a spiritual fire fanned by the reciprocities of others.

For Brother Mullins and his helpers, these Sunday-afternoon broadcasts are highly cathartic, leaving them drained and almost lethargic after thirty minutes of such intense expressiveness. Although it takes a few moments for this group to "come down" after a broadcast, they look quite tranquil by the time they have gathered together all their sound equipment and are ready to leave. "I don't intend to hold nothing back when I speak for Jesus," says Mullins.[18]

Serviceable to needs of the studio participants. Spirited programs such as the one described above often suggest that airwaves-of-Zion performers benefit more from the broadcasts than any listeners could. Not only do these preachers and singers exalt in the emotional scenes experienced, but they claim personal benefits arising from the fulfillment of their evangelistic missions. Indeed, this sense of mission tends to grow much stronger the longer an airwaves-of-Zion involvement lasts, and it is one of the "common threads" I discuss in the closing chapter of this volume. "It's been my life," says Sister Dollie Shirley, speaking of her forty-one years with "The Morning Star Gospel Program."[19]

Commitment to a mission seems particularly intense in the case of the longtime lone exhorter, that individual who—unaccompanied by musicians and singers—travels each Sunday to an airwaves-of-Zion station to broadcast fifteen or thirty minutes of impassioned "witnessing." In chapter 4 we will examine Brother James H. Kelly, a radio preacher for almost forty years and my prototype of the lone exhorter who week after week makes that pilgrimage to a microphone, in this case at WNKY, Neon, Kentucky. Kelly, as we will see, comes to WNKY with nothing more than his well-worn Bible, takes his position in front of the WNKY mike, and preaches an improvised sermon for thirty minutes, continuing a task he says God gave him to perform four decades ago. When I asked him if he felt confident that people listened to him, his only response was, "I get a letter now and then."[20]

Brother Kelly, however, is only one member of the lone-exhorter type I have encountered elsewhere. The following preachers can be named as examples: Brother David Barnette, WAEY, Princeton, West Virginia; Brothers Ronnie McKenzie and Dewey Russ, WELC, Welch, West Virginia; Sister Sue Hill, WBPA, Elkhorn City, Kentucky; Sister

Ramona Coles, WJLS, Beckley, West Virginia; and Sister Myrtle Lester, WNRG, Grundy, Virginia.

Like Brother Kelly, Sister Myrtle Lester has been on the air for what she stated simply was a "number of years." In addition, she exhibits Kelly's determination to keep proclaiming an evangelistic message, apparently for as long as she can find a mike and is physically able to preach. In a delivery that for the airwaves-of-Zion genre is uncharacteristically low in volume and dynamism, she pleads with her audiences to accept a basic evangelical thesis of sin, repentance, grace, regeneration, and salvation. Although speaking in a voice that shows the energy-draining consequences of age, she still communicates an intensely devout determination to "witness." Her voice trembles with emotion when she says, "God reaches out to those that will come to him and repent and give him their hearts and give him their lives. The saddest thing there is is for a soul to go out of this life unprepared to meet God."[21]

Both Kelly and Lester need their weekly times before a microphone, and they both would feel unfulfilled should the respective stations discontinue airwaves-of-Zion programming. Although they would define their respective missions in terms of some drive to evangelize, it appears obvious that they have their own personal injunctions to follow. Still, even in Kelly and Lester, I have never seen this "need" manifested more forcefully than in Sister Dollie Shirley, a need that I will attempt to capture in its greatest poignancy when, later in this chapter, I discuss the closing out of that forty-one-year-old "Morning Star Gospel Program" tradition. Furthermore, we will see that Sister Dollie did not stand alone in experiencing that need.

Not comparable to televangelism. There are few comparisons between the big-money, high-tech world of televangelism and the world of the airwaves of Zion. In fact the only similarities might be that both deal with religion and do so—generally speaking—at a highly emotional level. The airwaves-of-Zion movement has had no access to expansive markets, domestic or foreign; it has raised no huge sums of money, except in some very broad collective sense; it has created no "super-preachers"; it has established no educational institutions, hospitals, broadcasting systems, or extensive missionary networks; it has instituted no publishing houses or other large-scale commercial operations; it has played no highly visible role in promoting a religious-right political agenda, even though it does support many of the individual positions taken by conservative groups; it has not been accused of stealing followers, finances, and other forms of support from local churches, a charge that has become controversial even for the larger electronic church;[22] and it has not given rise—or is it likely to do so—to any scandals that have swept the nation.

Concerning the question of money appeals—a subject I promised to mention—it should be noted again that the typical airwaves-of-Zion preacher makes few highly overt and formal requests for financial contributions, if any. Most of these airwaves-of-Zion preachers are like Brother Roscoe Greene, urging listeners to "write in," but never specifically mentioning the need for money. If contributions do come in, individual givers will be thanked without saying anything about amounts; and rarely will one of these preachers call for a named amount of money needed to stay on the air. Sister Ramona Coles, an African-American Holiness preacher broadcasting over WJLS in Beckley, West Virginia, even tells her listeners, "Don't you send me no money. You need that. You send me your letters of support." She gets some money anyway.[23]

Solicitation of money is almost always an awkward practice for Appalachian preachers, imbued as they are with the principles of an unpaid ministry and with the thesis that they should support their spiritual labors with the return from their secular labors. Still, airtime does cost, and they are forced to rely on the small contributions of their listeners. In spite of this, however, it's rare to hear an airwaves-of-Zion preacher make a direct call for dollars.

I will discuss examples of begging for money in chapter 5 when I examine a tent revival in which Sister Brenda Blankenship participated, and to some degree in chapter 2 when reporting Brother Johnny Ward's evangelistic aspirations; but to a large degree the radio programs I observed during this study were free of those droning-on-and-on requests for financial contributions common to televangelism. Furthermore, the airwaves-of-Zion contributions that are made are apparently small ones, generally ranging from one to ten dollars.[24] One exception to this general rule is mentioned later in this chapter, and another is detailed in chapter 4.

Of course the airwaves-of-Zion preacher is operating over a medium far less expensive to access than that over which the televangelist is heard; nevertheless, I am convinced that there are considerable differences between the value bases of these two electronic church phenomena. Among airwaves-of-Zion exhorters, I sense a genuine reluctance to talk about money, on the air or off, as if such discussions would detract significantly from their evangelistic missions.

Typically, airwaves-of-Zion preachers pay their airtime fees with out-of-pocket money, home-church contributions, and donation receipts, with the latter admittedly being relied upon most heavily. However, my conversations with these persons generally reveal—if my sources can be trusted—that contributions stay about even with airtime costs, suggesting that no great sums are garnered by the program's principals. Throughout his tenure on WATA, Brother Roscoe Greene's mode of operation was to come to the station each week (on a weekday)—dressed in his overalls, the station's staff say—pick up such correspondence as had come to WATA in his name, pour all metal or paper currency onto a desk for counting, endorse all checks to the station, and in the process pay his airtime costs several weeks in advance. This routine convinced station personnel that all the money Greene received went back into the program.[25]

I will not argue that profits are never made from airwaves-of-Zion broadcasts, but I do contend that such financial gains are marginal, probably never sufficient to pay a minimal wage for time involved, especially if one considers all of the people in any particular program. Suffice it to say that these individuals appear driven by something other than money.

Falls under the "folk religion" heading. The term "folk religion" is problematic, having been variously defined by scholars,[26] but I employ it here in what I think are the term's traditional applications to regional study and in a sense that is close to, if not identical with, meanings given the phrase by folklorist Elaine J. Lawless.[27] Thus my use of "folk religion" means that I view the airwaves of Zion as a regional common man's phenomenon, originating from heavily independent religious subgroups that owe little or no allegiance to hierarchical structures above them. These subgroups express their beliefs and communicate their passions through a regional vernacular that is often foreign to mainline denominations, and they hold doctrines strongly rooted in oral traditions, employing stylized sermonic techniques that in many cases are distinctly Appalachian. They are oriented more toward intensity in spiritual experience than to doctrinal exactitude and are usually led by exhorters devoid of formal training in theology or homiletics. Their audiences are generally below the national average in educational and other socioeconomic measures. If the term "folk religion" engenders, for any reader, images of non-Christian, primitive, tribal spiritual expression, then that individual may wish to employ a substitute phrase like "religious expression of a regional folk culture," words that communicate more explicitly my meaning.

Addressing southern religious traditions in general, David Edwin Harrell, Jr., uses the term "plain-folk religion," not precisely as a synonym for "folk religion," but as a phrase that denotes a religious "class" apart from those more establishment-oriented mainline groups.[28] Add the regional factor and the airwaves-of-Zion broadcast genre certainly fits within that designation, so much so that the typical mainline worshiper probably at best finds Roscoe Greene, Garrett Mullins, Brenda Blankenship, Dean Fields, Johnny Ward, and others mentioned in this volume to be backward, quaint, and at least mildly embarrassing. At worst, these mainliners (and here I include perhaps those modern evangelicals who have built movements that are often national and international in scope) consider the Brothers and Sisters of airwaves-of-Zion programming to be deplorably primitive and frequently heretical. While an Assembly of God congregation might find more identification with airwaves-of-Zion theology than would a Lutheran, Presbyterian, or even Southern Baptist congregation, none of these groups would be completely comfortable with the totality of this phenomenon, finding its people, programs, and principles unacceptably untutored and ungoverned.

The lingering presence of the phenomenon in Appalachia. Nowhere in this volume do I claim the airwaves-of-Zion phenomenon to be exclusively Appalachian; nevertheless, the earlier isolation of this region has caused many traditions to linger longer in these mountains than elsewhere, and that appears to be true relative to this particular tradition. During the last several years, as I have traveled throughout central Appalachia, engaged in the fieldwork this volume required, I have frequently found myself discussing the project with various individuals in and out of the airwaves-of-Zion camp. Invariably the person with whom I would be talking would say something like the following: "Have you visited station _____? They've got Brother _____ (or Sister _____) on the air there, and have been airing that broadcast for _____ years or so."

Suffice it to say that in Appalachia I have been faced with no dearth of examples to study. Indeed, my challenge was to select stations and broadcasts that seemed truly representative of the phenomenon. In the process I was forced to pass over several appealing suggestions for study, long-running, colorful, and ethnographically interesting broadcasts over such stations as WLSD, Big Stone Gap, Virginia; WLSI, Pikeville, Kentucky; WBEJ, Elizabethton, Tennessee; and WBBI, Abingdon, Virginia; while also giving only cursory attention to several stations I did visit, such as WNRG, Grundy, Virginia;

22

WLRV, Lebanon, Virginia; WDIC, Clinchco, Virginia; WBPA, Elk-horn City, Kentucky; WKSK, West Jefferson, North Carolina; and WJLS, Beckley, West Virginia.

My initial 1989 telephone survey of central Appalachian AM broadcasting units did turn up some stations that had only recently (within the last five years) canceled all of their airwaves-of-Zion programs—WHJC, Matewan, West Virginia, for example. This dis-covery suggests that Appalachia may be entering a transitional stage during which this genre of radio programming will gradually dis-appear. That observation leads me to my discussion of the final airwaves-of-Zion defining characteristic.

A fading phenomenon. Like many of the more traditional religious practices of Appalachia, the airwaves of Zion gives every indication of being in decline—in terms of participating radio stations, actual number of programs originating from these stations, and body of lis-teners. In short, the phenomenon may be dying, and that fact alone has played a major role in motivating this study.

This genre of broadcasting is threatened, in part, by changes oc-curring in that segment of the communication industry to which it is attached. First, during the 1970s and 1980s, AM radio in general steadily lost ground to FM, dropping from a 75 percent share of the total radio audience market in 1972 to only a 24 percent share by 1988,[29] this in spite of the top two or three stations in many markets still being on the AM band. This precipitous decline in the AM audi-ence caused considerable concern from the mid-1980s through the present for what might be done to save the AM band.[30] Second, the AM audience that has remained with airwaves-of-Zion stations has become an older, less affluent, more rural demographic group, who in turn is less attractive to advertisers than that younger, more up-wardly mobile, freer spending FM-listener group, particularly when the FM audience is compared with that listener segment still held by AM "stand-alones" (those AM stations having no sister FM station with which to engage in simulcasting).

Struggling for both listeners and advertising dollars—the former of course determining the latter—AM stations often have been forced to make drastic changes in their programming: through "narrow-casting"—shifting to highly restricted formats such as all-Elvis, all-children's, all-business, all-sports, or all-religious programming; or through automation—dropping live broadcasting and relying solely on prerecorded (often syndicated) segments of music, talk, and enter-taining news commentary (from Paul Harvey and the like), plus the more up-to-the-moment satellite-transmitted syndications of news and weather.[31] Add to all of the above the 1982 FCC deregulation removing requirements to report a station's percentage of public service programming and announcements, subsequently reducing the incentive for both AM and FM stations to air local religious broadcasts as part of their public service programming. Prior to this deregulation, some stations gave free time to local religious groups.

All of these industry adjustments and governmental deregulation have produced an environment generally inhospitable to the airwaves-of-Zion phenomenon, and this attitude is present even in the narrow-casting situations that result in all-religious formats. In the latter circumstances, stations often buy into syndicated programming taped by such production and distribution companies as Broadcast Programming Incorporated, which currently offers four separate "Christian Formats," or the Christian Broadcasting Network, which at the time of this writing has 240 affiliates.[32] The typical airwaves-of-Zion broadcast would be incompatible with the much more polished programming distributed by these syndicators.

Also at the time of this writing, new policies are being considered and new technologies being developed and promoted that promise to improve both AM transmission and reception—an expanded AM band, the reduction of AM interference, digital audio broadcasting (DAB), consumer use of digital receivers, and others[33]—but these new policies and technologies threaten to move AM further away from any local-programming base, thus posing further problems for these airwaves-of-Zion broadcasts. In addition, since so many small AM stations lost money during the late 1980s, there appears to be an immediate danger that market forces will eliminate many of the very stations over which airwaves-of-Zion programming is aired.

In numerous cases, the communities in which these small AM stations are located also have experienced drastic culture-base changes that in turn influence radio-listening tastes: (1) in-migrations brought about by tourism, second-home real estate markets, industrial and commercial growth, the return of earlier out-migrants, and the occasional presence of a thriving educational institution; (2) the ordinary changes generated by improved education and access to contemporary communication technology; and (3) the various involvements of both private and public agencies for social and economic change.

The Blue Ridge Parkway, the Great Smoky Mountain National Park, the New River Gorge National River, the numerous state parks, a large number of scenic lakes and white-water rivers, and a host of privately built resorts or theme parks such as Grandfather Mountain in North Carolina and Dollywood in Tennessee constitute only a part of the total collection of attractions that bring visitors to southern and central Appalachia; and these sojourners have played a role in reshaping the cultural base of the region. Furthermore, the new industry and commerce mentioned above have brought to some regions of the southern mountains managers, business owners, their families, and an entire cadre of population imports that have gradually contributed to the cultural changes in the region.

WATA-AM and "The Morning Star Gospel Program"

The transition that has occurred at WATA-AM, Boone, North Carolina, can be used as an example for much of what was discussed above. When this station first went on the air in September 1950, broadcasting from makeshift studios over the old bus depot, it served an audience composed principally of indigenous residents of Watauga County and the students and faculty of Appalachian State Teachers' College. That audience now includes (or has the potential to include), in addition to the "locals," heavy segments of tourists (all seasons of the year), seasonal ski-resort visitors, seasonal second-home residents, a large body of professionals and businesspeople, an ever-growing retirement community, and the greatly expanded and vastly changed student body and faculty/administration of Appalachian State University.

These changes in WATA's market have in turn precipitated subtle and not-so-subtle alterations in the station's style and programming. One example of this can be seen in WATA's early-morning broadcasting. For most of the 1970s these initial hours of the station's day were controlled by an announcer who, as a part-time minister, part-time policeman, and part-time radio personality, developed considerable rapport with listeners indigenous to Watauga County, North Carolina, creating a station/audience relationship that was direct, intimate, folksy, and personal. During these beginning-of-the-day broadcast hours, this announcer would call people around the county and engage them in over-the-air chats, in some cases finding out what the individuals were having for breakfast, the progress of crops in a particular mountain valley, or—during the wintertime—the condition of roads. Listeners also felt free to phone this announcer, thus maintaining an open relationship between the station and its out-in-the-county audience.

The only problem with this approach was that a growing segment of WATA's listening audience began to find the chats far too folksy, banal, and colloquial, playing almost exclusively to a rural audience. The university students, in particular, chose not to listen, and this was a body of consumers the station's advertisers needed to reach.

Sensing this shift in listener attitude, the station gradually moved away from that more personal and chatty style of audience contact, particularly as this chattiness applied to rural listeners. In the process, the station abandoned a mode of operation that has meshed so well with the more rustic intimacy of airwaves-of-Zion programming.

Although WATA cannot be considered representative of the airwaves-of-Zion affiliates examined in this work, it does exemplify a station that has been heavily influenced by the economic and social transitions mentioned above. Twenty years ago WATA was an airwaves-of-Zion station, including in its Sunday programming a heavy concentration of the types of religious broadcasts in this study. Indeed, Brother Roscoe Greene's "Morning Star Gospel Program," which aired over this station for almost forty-one years, still stands as my prototype for the airwaves-of-Zion genre.[34]

When I wrote the first draft of this chapter I thought "The Morning Star Gospel Program" would continue for several years to come, perhaps reaching a forty-five- or even fifty-year mark. I was prepared to call WATA a "transitional station," in the sense that it was continuing to broadcast only one airwaves-of-Zion program, with the probable future of moving completely away from the genre once the aging principals of that show retired.

That scenario, however, developed a little sooner than I expected. On August 4, 1991, Brother Greene announced to his "Morning Star" audience that the following Sunday (August 11, 1991) would become the last airing for the program. He had been ill, he said, and so had his wife, Sister Hazel, who regularly made all of the show's announcements. They could not continue, he tearfully added, to meet their obligations to the broadcast.

I was in the WATA studios that August morning, and again the following Sunday. When his announcement was made, Brother Greene was not the only person who cried. The sixteen people who were a part of that next-to-last broadcast of "The Morning Star Gospel Program" all wept, some uncontrollably. Sister Dollie Shirley, the only person who had been with the program since its inception, was particularly distraught, for a while unable to speak or sing.

It was Dollie Shirley's father, Brother Bob Smith, who instituted

the broadcast in 1950 (the first Sunday WATA was on the air), building what was then a fifteen-minute program around the singing of the
Morning Star Trio—Brother Smith, Sister Dollie, and Brother Stewart Hamby. On the death of her father, Sister Dollie became the
leader of the program's singers, which remained a trio for a while and
then became, until the show's close, whatever combination of musical
talent Shirley could pull together. Over those forty-one years of
broadcasting the singers and musicians Sister Dollie brought to
WATA became the heart and soul of this airwaves-of-Zion program,
aided of course by Brother Roscoe Greene and Sister Hazel Greene;
moreover, it was Sister Dollie who really kept the tradition going, especially during two periods when Brother Greene briefly left the
broadcast. Throughout this lengthy tenure of "The Morning Star Gospel Program"—well over two thousand broadcasts—Shirley failed to
perform her program duties only seven times, all on Sundays when
either she was immobilized by illness, or the roads of Watauga
County, North Carolina, were so buried in snow that travel to the station was impossible. Thus her choked comment, "It's been my life," to
me seemed justified in scope and tone, expressing a degree of commitment that would be difficult to dismiss with any "So what?" shrug of
either apathy or cynicism.

Brother Greene might have been as vital to that continued tradition had he not pastored during those years as many as three
churches at a time, meanwhile staying in constant demand in a mul-
ticounty area as a revival evangelist. He learned to rely on Sister
Dollie, knowing that she would always prevail in the task of finding
someone to preach whenever he was unavailable.

On August 11, 1991, the close-out performance of "The Morning
Star Gospel Program" became even more intensely emotional than it
had been the week before. Some members of earlier program casts returned for this final production, and much of the airtime was devoted
to reminiscing, with several of the participants saying, and then
repeating to themselves, "The next time we'll all meet will be in
heaven."

At times the broadcast itself became almost unintelligible, as a result of all the crying, embracing, and general movement throughout
the small studio. I snapped my pictures of the event and experienced
my own emotions, in empathy with this group of dedicated people who
were about to end something that had meant so much to them.

During this run of almost forty-one years, the program's theme had
been the old hymn "Must Jesus Bear the Cross Alone?" The broadcast

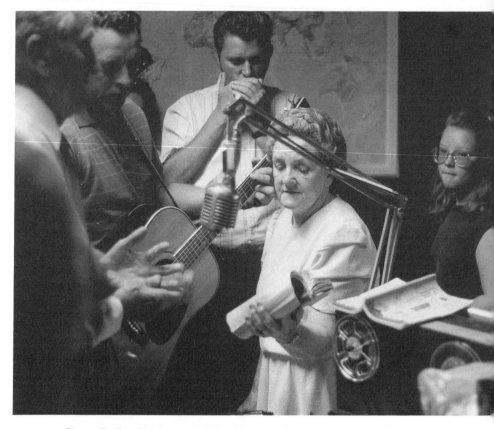

Sister Dollie Shirley and "The Morning Star Gospel Program." The final broadcast.

had always opened and closed with the singer's rendition of this be-
seeching question. On that final morning Brother Roscoe opened his
remarks by asking the group to "sing it again."

> That song's gonna have to be sung again. Right now, Amen!
> I want to say a word to you before they sing this song again. I want
> you if you don't get anything this morning out of the Morning Star
> broadcast, I want you to get the message from this song. It's been sung
> here, if we've got the record correct, about 2,184 times. That song's been
> sung here on the Morning Star broadcast. I want you to get it. This song
> has inspired me every time that I've ever hear'd it sung. It's encouraged
> me. I mean it's absolutely encouraged my soul when they sang that
> song. Listen at the words of it now before we pray. Listen. All right,
> Dollie.

What then followed was Sister Dollie and the singers' rendition of
the first, third, and last stanzas of Thomas Shepherd's well-known
hymn:

> Must Jesus bear the cross alone,
>   And all the world go free?
> No, there's a cross for everyone,
>   And there's a cross for me.
> The consecrated cross I'll bear,
>   Till death shall set me free;
> And then go home my crown to wear,
>   For there's a crown for me.
> Oh, precious cross; oh, glorious crown;
>   Oh, resurrection day;
> Ye angels from the stars come down
>   And bear my soul away.

Brother Greene was a little off on his count of the number of broadcasts of the show: the total appears to have been only 2,132, minus
a snowed-out broadcast or two. The error in his count came from
Greene's thinking they were approaching their forty-second year on
the air rather than their forty-first. Either way, the group's accom-
plishment should be considered significant, an achievement that
could only be obtained by an exceptional degree of dedication and per-
severance.

This second singing of "Must Jesus Bear the Cross Alone" was cli-
mactically emotional for Brother Greene, and he interjected numerous joyous exclamations or directives: "Listen there. Amen!"
"What a joy!" "Listen, my friends. The precious cross!" "The resurrec-

tion day. That's the morning!" Throughout this process of crying out with exultant phrases, Greene seemed to be reaching for some final verbalization of joy—some ultimate expression of meaning—a summit statement that would provide the capstone for forty-one years of effort.

In general his words did not capture that expression nearly as forcefully as did the nonverbal behavior of this group of celebrants, absorbed as they were in their individual statements of elation and pride. I watched a young girl, of early-high-school age, who had been with the program only a couple of years, a daughter of one of the older program participants. For those two years her contribution to the broadcast had been through her rhythmical playing of a pair of spoons, but through those efforts she had become a "Morning Star Gospel Program" cast member, fully eligible now to share in this group emotion.

She cast her eyes downward, not wanting me to photograph her act of crying; and her hands trembled as she made her now imperfect attempts to stay with the rhythm of the hymn, finally giving up this effort as her father, the group's harmonica player, pulled her into his arms.

When the emotional reactions to that second singing subsided, Brother Greene reminisced about the group's own effort to "carry the cross," talking in part about the three places WATA had been housed during those years—the second floor of the bus depot on Depot Street, the second floor of the Watauga Savings and Loan building on King Street ("up town"), and the relatively new studios on Highway 321 ("down here")—and about one woman who had "sponsored" the broadcast during its entire tenure:

> You know I feel just like lifting my hands and praising the Lord this morning, Amen, for all of his wonderful love he has bestowed upon us all these many years.
>
> I wonder how many that's still carrying the cross that started when we was up yonder over the old depot. You know my mind runs back to all those mornings we met up there over the old depot, and went on up town, you know. Up there. I don't know how many years we was up there and then moved down here. And I just wonder—I wish I knew this morning—how many that's faithfully carried the cross all these forty-two years.
>
> Now we've got one Sister that's been very faithful. I know we've got many that's faithful, and I appreciate everything you've done, every

prayer you've prayed. Everything that you've done I appreciate it from the very depth of my soul.

Sister Cleo Tester, I guess she's the oldest sponsor that we have on the Morning Star broadcast, and most of the time, if my mind serves me correctly, since she's started she's paid thirty dollars a month on this broadcast for all these years.[35]

The following week I phoned Sister Cleo Tester to talk to her about those years as a "sponsor" and to gain some understanding of her commitment to the broadcast. That commitment was strong, and she said over and over again, "I'm gonna miss Preacher Ros."

She told me about her own visits to the program, particularly during the 1950s when the station was over "the old bus depot." "Seemed like he was such a strong preacher in those years," she said. "And the singers were so good. I kept up with all the people who were on the program, but I've forgotten some of them now."[36] She also told me that "Preacher Ros" had never asked her for her money contributions, but she added that she did appreciate her name being mentioned occasionally by "Sister Hazel."

Sister Cleo Tester was unable to provide the more exact details I wanted about those early years: her own memory had faded. However, she did communicate the depth of her feeling for this tradition that had now ended. In doing so she told me something about age, and memories, and the universal need to hold on to part of yesterday.

During that final broadcast, Greene went on to thank all the singers who had been with the show over the years, speaking very personally of some of them. On one name he stalled, not being able to remember it. Sister Hazel helped him out. "See, I'm getting old," he said. "I'm not complaining, but since I've had high blood it's in some way affected my memory."[37]

He also expressed his disappointment that the show was having to close. Without being specific, he alluded to someone whom he had hoped would take up his mission and continue the program. A preacher apparently had told him that he would do just that, but the individual and his supporters had changed their minds. This narrative precipitated Brother Greene's only negative comment: "They proved unfaithful. I'm gonna say it from the bottom of my heart. They proved unfaithful."

It should be noted here, however, that the station management had become determined to close out the program once one or more of its key participants died or retired from the broadcast. Such a realization may have influenced the decision of the individual or individuals to whom Greene was referring. Indeed, it may have mandated it.

This was a rare moment of direct, personal, and somewhat bitter criticism from Brother Roscoe Greene. Regular listeners to "The Morning Star Program" seldom heard strongly negative remarks from this man, but now he was being compelled to end a tradition that meant much to him. He had apparently hoped that the institution would not die, and he was experiencing the pain of reality.

"The Morning Star Gospel Program" did not end on that low note. By the close of this final broadcast the two dozen people who were present celebrated joyfully the forty-one-year existence of this thing they had created, clinging to each other to share emotions of exaltation and sadness. Their creation was about to die, but they felt good about all the effort that had earlier kept it viable.

It was better that "The Morning Star Gospel Program" close this way than suffer some more ignominious demise. Jim Jernigan, current program director of WATA, had discussed with me the problems involved in keeping this type of programming on the air. Jernigan had mentioned in some detail his difficulty in meeting the needs of WATA's diverse audience and had observed—weeks before Brother Roscoe Greene made his decision—that when any of the key participants in the "Morning Star" broadcast ceased their respective involvements with the show, the program would, "in all likelihood," be ended, thus bringing to a close the airwaves-of-Zion era of this particular station.[38]

My main point here is that WATA has passed through a transitional stage, a stage eventually to be reached by the vast majority of central Appalachian AM stations, although not as a result of exactly the same set of dynamics. When such transitional stages are reached, station owners and/or program managers often become embarrassed by their airwaves-of-Zion offerings and seek ways to phase out that type of programming. Indeed, on three occasions during the fieldwork for this study I was told by station management personnel that they hoped I would not suggest in my writing that the airwaves-of-Zion program under study was "representative" of the respective broadcasting unit. They wanted me also to look at their regular weekday programming.

Summary

The forecast for airwaves-of-Zion programming looks dark. At some time in the future the "Brothers" and "Sisters" who produce such Sunday offerings as "Gospel Echoes," "The Old Fashioned Gospel Time," "Airwaves of Grace," "Voices of Zion," and "Sounds of Salvation" will find studio doors closing to them, and the practice of this particular folk religion tradition will end. In the place of these programs, listeners will undoubtedly hear other religious broadcasts, but these new productions will exhibit technical sophistication, contents, and formats not currently present in the airwaves-of-Zion phenomenon. In addition, the improvisational quality, the first-person intimacy, the rustic modes of speech, the immediacy of listener response, and perhaps some of the honest passion will be gone.

All of this will not happen immediately. Undoubtedly there will be stations that preserve this broadcast genre long after the main corpus of the phenomenon is dead, but for the typical Brother Roscoe Greene and Sister Dollie Shirley the AM radio environment will have changed, making it much harder for them to find an airwaves-of-Zion studio from which they can send forth their "plain folk" evangelism. When that happens Appalachia's AM airwaves may become less culturally varied and rich by the loss.

The Case Studies

The remainder of this volume consists of four case studies, each of which focuses, first, on the respective Appalachian region; second, on one airwaves-of-Zion station; third, on that station's live Sunday religious programming; and, fourth, on one of those programs (but not necessarily in that order). In addition, as a part of each case study, I give considerable space to at least one outside activity of the respective individual or group, attempting through this effort to "round out" the descriptive ethnography, fitting the subject into a larger folk-religion context. A final chapter is devoted to integration and analysis.

The choice of these particular stations and programs emerged, I confess, less out of design than out of a complex set of serendipitous happenings. Nevertheless, with stations WMCT, Mountain City, Tennessee; WAEY, Princeton, West Virginia; WNKY, Neon, Kentucky; and WELC, Welch, West Virginia, we have a representative spectrum of the transition I suggest is taking place within the airwaves-of-Zion phenomenon.

WNKY, in Letcher County, Kentucky, is still solidly fixed within the movement and shows few indications of a transition to a programming philosophy hostile to airwaves-of-Zion broadcasting. This appears to be true in large part due to the relatively unchanging nature of WNKY's listener market: although that audience of listeners may have grown smaller and older, there is little indication that it has evolved to some decidedly different cultural base.

WMCT, Johnson County, Tennessee, on the other hand, may be poised for some shift away from its heavy Sunday concentration of airwaves-of-Zion programming. This potential for change might not be seen in the town's population history, which shows an increase from 1,379 in 1960 to only 2,169 in 1990. Instead, I base this speculation in part upon the sensitivity of the station's owner to my study, and her wanting it to be made clear that she did not consider the program I examined to be "representative" of her station.[39] In addition, at the time of this writing, the WMCT owner, Mrs. Fran Atkinson, is in competition for an FM license. If successful, this effort could alter the set of dynamics currently influencing WMCT's programming policies. Finally, recent economic and cultural changes in Johnson County, Tennessee—new industrial growth and the beginnings of a second-home real estate market—suggest the possibility of listener-market changes on the horizon.

WELC, McDowell County, West Virginia, has now acquired a sister FM affiliate and is AM/FM simulcasting during much of its broadcasting week—certain segments of its weekday schedule and most of its Sunday programming being relegated solely to its AM band. However, the socioeconomic base of McDowell County seems relatively stagnant at best and sharply regressive at worst. One result of this socioeconomic stagnancy is that Sam Sidote, owner/manager of WELC, envisions no significant divergence from his current programming policy, which includes the preservation of his airwaves-of-Zion broadcasts but, as stated above, with the concentration of those programs on his AM band.[40]

WAEY, Mercer County, West Virginia, appears to be signaling an intention to move away from the airwaves-of-Zion camp of radio stations. I base this judgment, first, on this station's having so tightly isolated its Sunday live religious programming on its AM frequency, simulcasting all other times of the week; second, on the station's treatment of its airwaves-of-Zion performers, detailed in chapter 3 of this work; and, third, on the way the station is currently marketing itself. Owned by the Betap Broadcasting Corporation, WAEY has recently strengthened its position in the southern West Virginia market through "the installation of [a] new transmitter, new antenna and the latest state-of-the-art Audio processing boosters" and now promotes itself as the number-one station "in Princeton, Bluefield and Mercer County, West Virginia" and as reaching sizable markets in McDowell, Raleigh, Summers, Monroe, Greenbrier, and Wyoming counties in West Virginia, as well as Tazewell, Wythe, Bland, Giles, Montgomery, and Carroll counties in Virginia.[41]

Suffice it to say that this station, through a wedding of its AM and FM programming, is aggressively going after the advertising dollars of southern West Virginia and southwestern Virginia. In that climate, airwaves-of-Zion programming may be of negative value, and WAEY may be envisioning a near-future date when its airwaves-of-Zion broadcasting becomes a "a thing of the past." Colorful as is the "Songs of Salvation" program (featured in chapter 3), WAEY may soon be unable to afford its continuance in the station's programming, thus setting the stage for a farewell performance similar to the one described for "The Morning Star Gospel Program."

The four programs and their respective personae also provide a picture of the diversity present in the airwaves-of-Zion phenomenon. Brother Johnny Ward of Triplett, North Carolina (chapter 2), represents the multitude of Appalachian radio preachers who dream of extending their ministries through the acquisition of a tent, a transporting vehicle, and all the lighting and sound equipment necessary to the work of a rural traveling evangelist. The prototype of Brother Ward's dream might be the "Miracle Crusades" of the "Bible Revival Gospel Ministries, Inc.," conducted by "Evangelist Bill Daniel," to whom the reader will be introduced in chapter 5 of this work.

In contrast, Rex and Eleanor Parker of Lerona, West Virginia (chapter 3), represent the hundreds of dedicated individuals who serve the airwaves-of-Zion cause through music rather than preaching. "Songs of Salvation," the Parkers' weekly broadcast, also provides an example of contemporary commercial religious programming that has maintained stylistic ties to the golden age of radio, in this case station WHIS, Bluefield, West Virginia, and the early 1940s.

I selected for study the airwaves-of-Zion contributions of Brother Dean Fields of WNKY, Neon, Kentucky (chapter 4), primarily because his weekly broadcast becomes the product of one complete church fellowship, Thornton Freewill Baptist Church, situated within one of the several "closed down" coal-mining regions of Letcher County. Fields and his congregation also serve as excellent examples of a type of support community that occasionally develops around the activities of an airwaves-of-Zion production.

Finally, I chose to focus on Sister Brenda Blankenship of WELC, Welch, West Virginia (chapter 5), because she provides an example of the strong female exhorters who find their place in this airwaves-of-Zion world. Indeed, the Sunday programming of station WELC-AM will become my case study of that larger contribution of women to the phenomenon.

—2—

Brother Johnny Ward and "The Voice of the Word"

> Be patient therefore, brethren, unto the coming of the Lord. Behold,
> the husbandman waiteth for the precious fruit of the earth, and hath
> long patience for it, until he receive the early and latter rain.
>
> James 5:7

WMCT, Mountain City, Tennessee, was one of the first Appalachian
radio stations I visited back in the early 1970s. That original visit occurred two years after my arrival in Boone, North Carolina, in August
1971.[1] I began to catch broadcasts over WMCT as I drove north on
Highway 421 toward Bristol or Abingdon, and the programming and
the relaxed tone of this small AM station fascinated me, particularly
on Sundays when I would be exposed to so many colorful and spirited
religious groups. Also, this was the time—described in *Giving Glory
to God in Appalachia*[2]—when I was first becoming interested in the
highly rhythmical preaching of Appalachia. The WMCT preachers
employed a variety of Appalachian preaching styles and thus intensified that early interest.

The general programming I observed suggested that WMCT
served well its target market in rural and mountainous Johnson
County, 1990 population 13,766. This is the northeasternmost county
of Tennessee, a region of central Appalachia that, although economically low-scale, has managed to escape—in the past—many of those
cultural, ethnographic, and environmental anomalies, traumatiza-
tions, or downright disasters often precipitated in these southern
highlands, either by rapid and irresponsible industrialization or
rampant tacky and tasteless tourist-trade development. For example, the county contains no mining operations and none of those "Li'l
Abner Land" or "Hillbillyville" tourist attractions that play mer-

cilessly on negative stereotypes of the southern highlands. However, Johnson County does have several fireworks stands just across the state line, indicative of Tennessee's more liberal law relative to the marketing of this product than in Virginia and North Carolina.

One negative trade-off of this sparse industrialization and tourism is that the county has not experienced a substantial population growth in the last sixty years. By 1930 the population count was already 12,209. Following an up-and-down pattern, that county total struggled up to only 13,745 by 1980, compared to the 1990 figure reported above, 13,766. Measured in terms of these United States Census figures, Johnson County, Tennessee, has been relatively stagnant.

As judged by the census counts reported in chapter 1, the town of Mountain City appears to have fared no better than the county in its population growth, slightly less than an eight-hundred-person increase in forty years; but there are some indications that the 1990s may witness some changes for both the town and the county. New industry has only recently moved into the region, fresh growth in commercial establishments has occurred, and there are the beginnings of a second-home, mountain-retreat real estate market in certain areas of the county. As suggested in chapter 1, all of this may begin to alter the cultural base of Johnson County, Tennessee.

Joined in this high-country section of Tennessee by only a handful of much smaller communities (Trade, Midway, Shouns, Pandora, Doeville, Butler, Shady Valley, and Laurel Bloomery), Mountain City—as the seat of Johnson County—presides over a markedly provincial region of ridges, coves, valleys, and forests; fast-flowing trout streams; narrow serpentine state or county roads, plus one across-the-county and across-a-mountain two-lane federal highway (U.S. 421); hundreds of small farms, and numerous dairy or beef-stock agribusinesses; some logging operations; no heavy industry, but perhaps a dozen "clean" labor-intensive manufacturing plants; a segment of one man-made lake that has become ringed by marinas and a middle-class lake-front-property resort operation; one consolidated high school; and—important to this writing—perhaps two hundred churches, primarily fundamentalist Baptist groups (particularly Freewill and Missionary), independent Holiness-Pentecostal fellowships, Church of Christ and Church of God (the Cleveland, Tennessee, branch) congregations, Assembly of God affiliates, and representations of several of the mainline faiths.

The county is not idyllically bucolic, nor has it remained totally unaffected by environmental hazards or been preserved in a nineteenth-century mode. Nevertheless, the indigenous cultural base of the region has not been supplanted completely by in-migrants. Such changes in this cultural base as have occurred appear to have been generated more by slow evolution from within, as opposed to traumatizing alterations generated from without. Obviously, external influences have been felt in Johnson County, Tennessee, past and present. However, the comparisons I would make would be between this region and Avery County, North Carolina, at one extreme, so altered by the ski resort industry and an upper-class second-home real estate market that an indigenous base is hard to find; and with McDowell County, West Virginia, at another extreme, brutalized by a century of outside-owned coal-mining operations, stripped of much of its original beauty, and left to decay among abandoned networks of rusting mining structures. By contrast, Johnson County, Tennessee, appears to have been less radically influenced by external economic or social forces, with the negative payoff being some degree of stagnation.

The Station

Operating within a cultural environment exemplified by those hundreds of traditional and/or fundamentalist churches mentioned above, WMCT devotes its Sunday morning and afternoon programming exclusively to religious broadcasting, beginning with early-morning sessions of gospel music and church announcements, and followed by a mixture of prerecorded and live programs of preaching and singing. The live productions start as early as 8:30 or 9:00 in the morning and continue until late in the afternoon, usually ending between 3:30 and 5:00.

Currently, WMCT is definitely an airwaves-of-Zion station, with its Sunday programming possessing all the characteristics mentioned in chapter 1. As we will see, some "airwaves" preachers have found a welcome at this station that they could not find elsewhere. Nevertheless, there are those few signs, mentioned earlier, of the beginnings of a transitional process comparable to the one experienced by WATA in Boone, North Carolina.

WMCT's physical facility sits beside Route 19, about a half mile north of "downtown" Mountain City, in the direction of Damascus, Virginia. The station is housed in a small building that looks much like a two-bedroom house. However, the large letters *WMCT* displayed above the front entrance let travelers know what purpose the structure serves. In addition, on the left side (as one faces the building) is a down-link dish for receiving satellite-distributed syndicated programs.

"Studio A," the fifteen-by-fifteen-foot room from which live groups broadcast, is equipped with a lectern, one standing microphone, a long folding table upon which could be placed other mikes, a turntable unit, six chairs, and a late-thirties or early-forties console radio that appears to be nonfunctional. One wall is decorated by a handful of photographs of gospel singing groups (the Spensers, the Chuck Wagon Gang, the Oak Hill Quartet, etc.) and by the original logs for the first day the station was in operation, December 8, 1967.

During the twenty years I have followed WMCT's Sunday programming, there have been a variety of individuals working the main board, disc-jockeying the gospel music programs and handling such nonrecorded public service announcements and other on-the-air comments as are necessary. There have been several high-school students, at least one male senior citizen, and, at the time of this writing, Mrs. Mary Lou Hayworth, one of my favorite airwaves-of-Zion personalities.

Mrs. Hayworth, a woman in her mid to late fifties, has a decidedly "grandmotherly" image. She has had no formal training for radio. She just applied for this job and got it. The station manager actually wanted to hire her son, but he had decided to enter East Tennessee State University, in Johnson City, and would not be living in the area.

Anyway, Mrs. Hayworth asked if she could be considered for the job. The station taped her voice, made a decision that she would pass in that regard, and then trained her on the operations of the board. She caught on quickly and has been working the Sunday shift every since.[3]

Mary Lou Hayworth turns the dials and flips the switches in an unhurried, in-control-of-the-moment fashion, never appearing anxious about anything or rushed by deadlines of the moment. On the air she comes across in an informal and colloquial way, employing all the local idioms of eastern Tennessee and, in the process, establishing a firm identification between herself and her listeners. At the time of this writing she has been in this job for almost three years and is well settled into the routines of her work.

No description provided above should suggest images of a "hayseed station" run by a staff of semiliterate rubes, the "Hee-Haw" stereotype of rural southern Appalachian radio. That stereotype has been as unfair to the region as any of the other Li'l-Abner-land images of

Mary Lou Hayworth at the WMCT control board.

Appalachia frequently promoted in the popular media. Nevertheless, these descriptions should suggest a station in touch with the cultural base of its listeners, making the choices that strengthen or extend that identification. Mrs. Fran Atkinson, owner of WMCT, staunchly defends the "professional" image of the station, arguing that close attention is made to programming that best serves the region. She also expresses pride in what she labels "a decent station": "Children can listen to us."[4]

WMCT's Sunday Programming

Over these twenty years there also has been a significant turnover among the live performance groups, with Brother Douglas E. Shaw's "Bread of Life Broadcast" (eighteen years on the air) currently being the longest running of such productions at WMCT. In 1989 Brother Dwight Adams, a Baptist preacher, finally ended his connections with WMCT after over thirty years of Sunday broadcasts on this and other stations. I devoted some attention to Adams in *Giving Glory to God in Appalachia*.[5]

Even though the program titles and performing groups have changed during these years, a consistency in tone and style has prevailed, with the typical group being composed of three or four singers or instrumentalists, the preacher, and perhaps one or more individuals who simply participate as worshiping bystanders. Members of this last category of participants appear to be satisfied with this peripheral connection with the broadcasts, often content simply to stand inside the studio and observe. At other times, however, they quietly sing, pray, or become physically involved in the passions of a moment—arms raised, head back, eyes shut, silently mouthing a personal contribution, but stopping short of joining the audible sound that goes out over the airwaves of Zion.

The atmosphere that prevails for most of these programs is down-home, informal, and relaxed; unrehearsed and unstructured, except for the most basic of program forms; open to the improvisational expressions of the moment—"Spirit led," rather than "man led"; emotional to an extreme, but joyously so, for the most part; "Let everybody say a word for the Lord" participatory; and decidedly uncomplicated in theological messages, placing an emphasis upon spiritual celebration and exhortation rather than doctrine.

There are family groups, church-congregation groups, independent groups centered around one charismatic preacher, and occasionally the lone exhorter, controlling the mike for his or her thirty minutes a week, fervently evangelizing whatever small body of listeners might at that moment be tuned to WMCT, Mountain City, Tennessee. My use of "her" is deliberate, since Holiness-Pentecostal churches of the region do allow women preachers,[6] citing Acts 2:18, among other Bible verses, as their scriptural justification: "And on my servants and on my handmaidens I will pour out in those days of my Spirit; and they shall prophesy." Appalachian Baptist subdenominations, on the other hand, tolerate no female exhorters. During those twenty years that I have followed WMCT's Sunday programming, I can remember hearing only two female preachers. That count, unsupported though it is, lies in sharp contrast with the situation at WELC, Welch, West Virginia, where, as we will see later, women exhorters tend to dominate, if not in number then at least in dynamism.

Reminding the reader that these programs change frequently, I will now run down the list of live broadcasts a WMCT listener would have heard on July 22, 1990, one of several Sundays I spent at this Mountain City facility.[7] Some groups remain with the station only for a few months, while others, as exemplified by Brother Douglas Shaw's eighteen-year stay and Brother Dwight Adams's even longer commitment, are much more faithful.

At 8:45 that morning "A Look at the Word," conducted by Brother Jerry Hames, became the first live broadcast of the day. Hames, at that time, led the First Assembly of God Church in Mountain City, a "mission fellowship" that only recently had been established in the community. A young minister from outside the region, Hames did not fit the typical Appalachian-preacher mold. Indeed, there were the beginnings of a slickness in his style, suggesting that his role model might have been one of the several televangelist Assembly of God preachers.

At 9:00 A.M. Hames was followed by Brother Frank Woods and "The First Freewill Gospel Time," a thirty-minute show conducted by Woods and two other preachers. A number of Freewill (or Free Will) Baptist churches exist in Johnson County, with two of them within the city limits of Mountain City—South Side Free Will, to which I devoted some attention in *Giving Glory to God in Appalachia,*[8] and First Freewill Baptist, a much larger fellowship.

At one time First Freewill was affiliated with the John-Thomas Association of Freewill Baptists, an organization uniting approximately a hundred churches of the Appalachian and central Midwest regions.

But that connection was discontinued several years ago, with First Freewill becoming an independent fellowship.

South Side Free Will Church, which I visited on a number of occasions in the early 1980s, was at that time a highly traditional church, as demonstrated by a distinctly Appalachian-style of preaching and singing, its preservation of footwashing, and its wood-frame clapboard-sided meetinghouse; First Freewill is a much larger fellowship that conducts a more contemporary choir-assisted service, styled to the Nashville, Tennessee, gospel sound, and colored by an impassioned arm-swinging, torso-swaying, congregational response evangelism. When I observed him on July 22, 1990, Brother Woods's preaching also fit that set of metaphors—emotional, fundamentalist rhetoric expounded by an exhorter anchored to a gospel-music-album-cover image—carefully coiffured, sincere in tone but highly stylized in manner, exuberant but studied.

Between 9:30 and noon WMCT did not air any in-studio live productions. An hour of this time was taken up by a Mountain City mainline church service (First Baptist) broadcast over a remote hookup, and an additional half hour was filled by a prerecorded call-in show taped by an Asheville, North Carolina, minister, the crux of which was that he attempted to answer all callers' questions on the Bible. Although some of these queries were general in nature, requiring scriptural interpretations rather than detailed knowledge of texts, this man did demonstrate an information base that could have come only from a lifetime of Bible reading and memorization. I have often been impressed by the amount of chapter/verse scriptural knowledge these otherwise minimally educated preachers possess.

From 12:00 to 12:30 P.M. the WMCT listening audience heard evangelist Johnny Ward and his wife, Sadie, with their "Voice of the Word" broadcast. Ward and his family will receive considerable attention later in this chapter.

At 12:30, Ward was followed by Brother Donald Penland, a nondenominational evangelist from Burnsville, Yancey County, North Carolina, who accompanied his program with recorded gospel hymns sung by his wife. The interesting thing about Penland is that he drives approximately 130 miles round-trip each Sunday to have his thirty minutes of WMCT airtime, evidence of his dedication to his mission, but also evidence of the difficulty some small-town radio preachers face as they find fewer and fewer stations willing to devote Sunday schedules almost exclusively to live religious broadcasts. Brother Johnny Ward lives in the community of Triplett, east of

Boone, but each Sunday he drives twenty-six miles northwest of Boone to access the airwaves of WMCT. Ward tried to buy time on WATA in Boone but was turned away.[9]

As was mentioned earlier, Mrs. Fran Atkinson currently is in competition for an FM license. At the moment it is unclear what influence such an acquisition would have on her AM programming. There is the possibility, therefore, that in the future even WMCT will sharply alter the nature of its Sunday broadcasting, simulcasting on both AM and FM frequencies.[10]

At 1:00 Brother Douglas E. Shaw and the Redeeming Grace Trio moved into the studio at WMCT and held the microphones until 2:00. Shaw then drove to Abingdon, Virginia, for his 3:30 broadcast over WBBI, closing out there at 4:00. I have already noted that Shaw's program is presently the longest-running such production at WMCT.

The sixth live broadcast of the day, from 2:00 to 2:30, became "The Way of Life," produced by Brother Charles Fletcher, pastor of Dyson's Grove Missionary Baptist Church on Highway 197, south of Mountain City. This program also usually features the Fletcher Family, a gospel singing group composed of Brother Fletcher, his son, and his daughter-in-law. However, on the afternoon of July 22, 1990, this trio was represented only by recordings, since Fletcher's son and daughter-in-law were at that moment performing at a homecoming at Pleasant Grove Baptist Church in Zionville, Watauga County, North Carolina. The full trio had been scheduled to sing somewhere that evening, but when Fletcher started to announce that event he could not remember the name of the church. Later his wife phoned the station and provided that information.

"Gospel Mission Broadcast" became the seventh live program of the day (2:30–3:00), led by Brother Gene Ward, who described himself as an old-fashioned Missionary Baptist. Ward was quickly followed by another Brother Ward, in this case Dean Ward, also a Missionary Baptist. "Ward" is obviously a frequently encountered surname in this region of central Appalachia. Dean Ward's show, ending at 3:30 P.M., closed out WMCT's July 22, 1990, schedule of live religious broadcasts.

A high degree of tolerance and cooperation is demonstrated among these performing groups, not only at WMCT but in all the small Appalachian radio stations I have visited during the past several years. Frequently productions are tightly sequenced in the programming schedule, as was the case at WMCT on July 22, 1990, requiring a close coordination of actions as one group gathers instruments and leaves a studio while another enters and sets up. Invariably these encounters are warm, supportive, and tactile, with handshakes and embraces quickly exchanged. In addition, it is not at all unusual for musicians, singers, or audience members to stay around for involvements with one or more later productions, especially when a preacher says, "We don't have Brother and Sister So-And-So this morning because of illness. Could one or two of you Brothers or Sisters help us out by singing or playing?" Furthermore, when a group arrives at a station it is not unusual for those individuals to walk quietly into the studio to join the program then in progress, confirming not only the open relationships between groups but also the informal atmosphere that prevails in these production facilities.

Although the respective groups may be denominationally different—Freewill Baptists versus Pentecostals, for example—there is usually a strongly shared sense of purpose and an accompanying camaraderie. Appalachian denominations, subdenominations, and sects are often more accepting of doctrinally different Christians than are mainline denominations, particularly when a shared "win souls to Jesus" mission takes priority over theological differences. "We're not trying to separate," says Brother Johnny Ward. "We carry the good news."[11]

"The Voice of the Word"

I chose this broadcast for closer examination not because I view it as being representative of WMCT, the concern of Mrs. Fran Atkinson that I previously noted,[12] but because it's a family production and because the six people usually involved with the thirty-minute show demonstrate such an intense commitment to their respective missions. The following, therefore, is a case study not only of "The Voice of the Word" but also of the passionately held beliefs that drive this group of dedicated evangels.[13]

Each Sunday at noon Johnny Ward and his wife, Sadie, open their show with a hymn that Sadie wrote, "I Don't Know What You've Come to Do, but I've Come to Praise the Lord."

> I don't know what you've come to do;
> I've come to praise the Lord.
> I don't know what you've come to do;
> I've come to praise the Lord.

46

I don't know what you've come to do;
    I've come to praise the Lord.
Praise the Lord, praise the Lord, praise the Lord!
Jesus is my God;
    I'm going to praise His name.
Jesus is my God;
    I'm going to praise His name.
Jesus is my God;
    I'm going to praise His name.
Praise His name, praise His name, praise His name!
If your God is dead,
    Why don't you try mine.
If your God is dead,
    Why don't you try mine.
If your God is dead,
    Why don't you try mine.
He's alive, He's alive, He's alive!
I've been baptized in His name;
    I've never been the same.
I've been baptized in His name;
    I've never been the same.
I've been baptized in His name;
    I've never been the same.
Praise the Lord, praise the Lord, praise the Lord!
He filled me with the Holy Ghost;
    I spoke in other tongues.
He filled me with the Holy Ghost;
    I spoke in other tongues.
He filled me with the Holy Ghost;
    I spoke in other tongues.
Praise the Lord, praise the Lord, praise the Lord!

As is true for all the hymns sung during this broadcast, this song follows an uncomplicated, repetitious melody, and is rendered with high enthusiasm and ever-escalating volume, thus opening the program on a loud but joyous note. Suffice it to say that the broadcast not only begins on this exuberant level but stays there throughout its thirty-minute duration.

After Sister Sadie's hymn, Brother Johnny Ward then leads off with introductory remarks similar to the following, delivered with the full force of his voice and at a rate of two hundred words per minute. Throughout this opening, as is true for the remainder of the program, Sister Sadie accompanies her husband's exhortations with

Brother Johnny and Sadie Ward at WMCT, Mountain City, Tennessee.

spirited or somber chords from her table-top synthesizer, building appropriately at dramatic moments. A learned-by-experience synchronization always prevails in the on-the-air actions of this husband/wife pair, guided by unspoken understandings.

> Praise the Lord! It's good to be back with you today. I thank God for another opportunity to stand and lift up the name of Jesus. This is "The Voice of the Word" broadcast. My name is Evangelist Johnny Ward, and I thank God, amen, just to be able to stand here this day and proclaim the word of God in Spirit and in truth. Praise God!
>
> I'd like to say if you have a need, if you're sick in body, if you're depressed or oppressed by the Devil, I want you to call this number, 727–6701. Praise God!
>
> Amen, our God's alive and well, praise God! And I want you to know, praise God, that he'll set you free. Amen! Hallelujah! He'll heal your body. Praise God!
>
> And if you're out there and not born again, you need to be saved. I want you to know, amen, that he died on Calvary that you could have life, praise God. That you could have it more abundantly. Praise God! I come today, amen, to tell a lost and dying world, praise God, there's still hope in Jesus. Praise God![14]

The broadcast that follows adheres to a fairly traditional model: there are hymns to be sung; there are church announcements, both for some regular services and for special revivals or singings; there are announcements also about other radio broadcasts, particularly for Brother Dewey Ward's show over WETB in Johnson City, Tennessee; there are calls for listener prayer requests; there is a time near the end of the half hour when these phoned-in requests are acknowledged and prayers are offered for the special concerns of listeners; there are pleas for listener responses by telephone or letter; there is preaching, accompanied by much shouting; and there is a closing prayer, joined by everyone in the studio, and often backed by an emotional hymn.

What must be emphasized, however, is that this is a very intense half hour. The songs, the prayers, and the preaching evoke unbridled responses from members of the show's cast, and also from the occasional visitors. These responses take the form of shouting, audible exclamations, impromptu prayers, background testimonies, and a variety of other verbal and nonverbal communications, frequently all mixed together.

Sermons, especially, become climactic expressions of passion, with the exhorter occasionally abandoning the standing mike and moving toward individual revelers who at that moment might be experiencing personal raptures, with both speaker and respondent reinforced by each other's emotions, all backed up by the music from Sister Sadie's synthesizer.

> There's always been hope, and there'll always be hope, praise God. Amen! Why don't you come to Jesus, praise God! Amen! For today's the day of salvation.[15]

So explosive are some of these broadcasts that the precise words going out over the air are not always intelligible; the composite expression of this ardor, however, is unmistakably clear, sending forth along the airwaves of Zion a style of evangelistic exhortation that becomes about as intense as such communication can be.

The traditional pattern of these explosive moments is for Brother Ward to start off slowly and softly and then work in conjunction with his wife and her synthesizer gradually to build the sequence into a peak of impassioned expression, often energized by shouts from other people in the studio, and almost always accompanied by considerable physical animation. If Mary Lou Hayworth sticks a phone message into the studio during such a sequence, Ward will use that event as motivation to move up several degrees in intensity. Sister Sadie Ward will then follow his lead and allow her music to build proportionally. The larger program proceeds to develop around these segments that build to a climax, separated by hymns, prayers, testimonies, and announcements.

The following represents one of these brief modules. At the beginning it was slow and measured, but by the close Ward was back to his two-hundred-words-a-minute delivery-shouting, waving his arms, and sometimes jumping.

> I'm glad today, praise God, that I've been made whole by Jesus. Amen. I want you to know [*a slight rise in vocal emphasis and general animation*] that Jesus cares. Praise the Lord! Amen! I don't care that you might be out there, [*increasing the rate and the volume*] and you might be on drugs. Praise God! You might be an alcoholic. Whatever! But I'll tell you [*lifting the intensity another degree, the music building with him*] Jesus died for you! He made a way that we might have life and have it more abundant. Praise God! People's calling in [*moving now to a much higher level of emotion*]! Just got a phone call, praise God, that wants someone else added to the list to be prayed for. They need to be born again [*almost shouting*]! And I want you to know, praise God, I'm going to be praying. And I believe that the Holy Ghost [*with a shout*]

will begin to move upon these people. And I believe the Spirit will draw, praise God! Hallelujah! [*His arms rise above his head in that traditional evangelical posture of praise or religious submission, and his wife once again increases the volume of the music.*] It'll give them every opportunity to make it right with Jesus. Praise God![16]

Program Personnel

A complete depiction of this broadcast phenomenon should include a detailed examination of the participants. I will begin with Johnny Ward.

As noted earlier, at the time of this writing Brother Johnny Ward and his family live in the unincorporated community of Triplett, Watauga County, North Carolina, approximately ten miles southeast of Boone. They occupy a small wood-frame rented house in Simmons' Hollow, a tight little cul-de-sac that progresses up Simmons' Creek.

As indicated in chapter 1, Ward identifies himself as a "full gospel," nondenominational evangelist, with a leaning toward "Jesus-only Pentecostalism." He believes in a "second blessing" possession by the Holy Spirit, a sanctification that allegedly follows the initial experience of conversion and allows the Christian to manifest more intensified levels of spirituality through speaking in tongues (glossolalia), healing, "dancing in the Spirit," "swooning in the Spirit," "running in the Spirit," and a host of other "possessed by the Holy Ghost" behaviors that have been pejoratively characterized as actions of the "Holy Rollers." Nevertheless, Ward rejects those more extreme forms of Appalachian Holiness-Pentecostal expression, handling snakes and drinking poisonous liquids (Mark 16:17–18). Although numerous people think these latter practices characterize the mainstream of Appalachian folk religion, in actuality these particular methods of spiritual expression are quite rare within the region, limited as they have become to tightly circumscribed areas of eastern Tennessee, eastern Kentucky, and southern West Virginia. This circumscription of snake handling as worship is due, in part, to state prohibitions against the practice, with West Virginia being the only exception to that rule.

Ward explains his "Jesus-only" doctrine by saying that he believes "Jesus was completely God—all God," that he needs only to pray "in the name of Jesus," and that when he baptizes he does so only "in the name of Jesus [Acts 2:38]," as opposed to saying "in the name of the Father, Son, and Holy Spirit."[17]

Precise doctrine, however, is not all that important to Ward. He argues that debates over theology are often meaningless and harmful, dividing Christians unnecessarily. "I've met people who all they want to talk about is what they believe," he says. "And Jesus is lost in the back-and-forth arguments."[18]

Ward produces his "Voice of the Word" broadcast with the help of his wife, Sadie, and occasionally with the added assistance of his brother, Dewey, and Dewey's wife, Cindy. Another couple, Daniel and Stephanie Riddle, also are always in the studio when the brother and sister-in-law are present.

All three of these men—Johnny, at forty-one; Dewey, at thirty-three; and Daniel, at nineteen—are preachers, labeling themselves "evangelists" and professing calls to "prophesy," indicating that they claim to speak, when preaching, by divine inspiration. In each case they point to a specific moment in time when they heard "God's call" to this mission.

In the case of Johnny Ward, he points to December 12, 1982, as his conversion date, noting that it was not very long after that event that he began to receive indications of a "call." "I used to attend these home services at my older sister's house in Mount Pleasant [Cabarrus County, North Carolina], and I got to testifying and then to preaching. One night God told me I would be a prophet. After that I started preaching in a lot of Pentecostal churches."

Johnny Ward dropped out of high school but managed to finish that part of his education when he was in the army. That was near the close of the Vietnam War. Both his brother, Dewey, and Daniel Riddle completed only the ninth grade.

None of these men is regularly employed, calling evangelism their full-time missions. Dewey recently established a church in Elizabethton, Tennessee, "The Jesus Faith Center," but the fellowship has not grown large enough to pay him a regular salary. All three survive solely by the "love offerings" contributed by the small fellowships to which they preach, the five- and ten-dollar gifts sent by radio listeners, and the odd jobs they occasionally find. Johnny Ward tells of the largest church collection he ever received, $168. The wives also are not regularly employed, contenting themselves to be of service to their husbands' ministries.

Johnny and Sadie Ward have two children, Wesley and Melissa, four and a half and two and a half respectively, at the time of this writing. The family lives in a rented house near the head of Simmons' Hollow in Triplett, cutting their own wood for heat and surviving largely off the canning cellar that Sadie keeps stocked with the products of summer gardens. However, they raise no chickens or other food animals, so they are compelled to purchase their meats and dairy products in grocery stores, augmenting their meager cash supply with food stamps. To a significant degree this family survives by the generosity of others, individuals who fervently support the calling to which Johnny Ward is dedicated. One "born-again" physician in the Boone area occasionally helps Johnny by providing lawn, garden, and shrubbery work.

Dewey and Cindy Ward live in Elizabethton, Tennessee, the location of Dewey's small church. This also places him closer to Johnson City, Tennessee, where he produces a 2:00–2:30 P.M. broadcast over station WETB. Whenever they can, Dewey and Cindy help with the noon broadcast in Mountain City and then travel the short distance to Johnson City for their own show. In addition, Johnny and Dewey spell each other when revival commitments necessitate absences from WMCT or WETB.

Daniel and Stephanie Riddle live with Dewey and Cindy in Elizabethton, sharing shelter, food, and other necessities in a rather remarkable communal existence. Dewey owns an old tent that he transports from small community to small community for revivals. He met Daniel Riddle one day when the young man stopped to help assemble this tent. During a break in the labor, while the two sat in the cab of an old truck, Dewey asked his helper, "Are you saved?" When Daniel said "No," a prayer session developed in the truck that resulted not only in Daniel's "born-again" conversion but in his full commitment to Dewey and Johnny Ward's evangelistic activities—the two radio shows and the constant round of revivals, most of them at small, independent Pentecostal churches.

Cindy, Dewey Ward's wife, is thirty years old and childless. She has been told by a physician that she is infertile. Like Dewey, she was educated through the ninth grade, and like the other two women she devotes herself completely to the evangelism that drives the lives of these six people. Cindy's main contribution to the ministry is that she plays the piano and sings, but she also provides guitar accompaniment for her own singing or for the gospel sounds of any of the other five. All of the women "testify" occasionally, either during radio broadcasts or at church services. Frequently these testimonies become sermonettes, accompanied by hand clapping, shouting, crying, or any of a number of possible emotional responses. As noted earlier, Holiness-Pentecostal traditions allow women the full range of reli-

Brother Dewey Ward at WMCT.

gious expression and practice, including preaching, administering sacraments, and the pastoring of fellowships.

Stephanie Riddle has the most formal education of any of the women, having received a high-school diploma. After graduation she worked for several months as the manager of a fast-food restaurant before marrying Daniel. At that time she quit her job and devoted herself to his calling. "I could go back to that work," she says, "but I believe I have been called to help my husband."

"I worked for several years at a job making over nine dollars an hour," adds Johnny Ward, making the point that he, too, could work if he desired, "but my whole mission is different now." To Johnny Ward it is critically important to his claim of "call" that "the Lord watches over" him.

"We're reaching people that the regular churches won't accept," reasons Dewey, suggesting that being relatively indigent somehow helps in that mission. "We preach the gospel to the poor and outcast." "We always survive," remarks Sadie. "I guess God is watching over us."

A depiction of these six individuals as "lazy and shiftless," finding any excuse to live off a welfare dole, would be unjust: there is much more pride in each of them than that. Sadie Ward, for example, found it difficult to admit that she had to rely on food stamps.

Still it must be noted that these people have divorced themselves from the traditional nine-to-five work ethic, believing their commitment to this evangelistic mission will sustain them. Those whom God calls, they argue, God will protect; and what the deity fails to provide ultimately will not be needed. It is a principle with consequences the Wards' two children have no choice but to accept.

Johnny dreams of getting a tent of his own, a newer one than that owned by Dewey, and perhaps a truck to haul it in, a rig similar to one that will be discussed in chapter 5. He has located one such tent, but that tent alone will cost him five thousand dollars. At the time of this writing he has just started a campaign to raise that money, mailing the following letter to numerous individuals and church communities with whom he and Sadie have worked over the last few years:

Dear Brothers and Sisters:
Greetings in the name of the Lord! We pray that this letter finds each one of you doing well. The reason for writing this letter is we have a burden to win the lost and to preach more in the year 1991 than ever before. We feel the Lord has given us the opportunity to do this work for Him by giving us the opportunity to buy a tent. We are asking each one that wants to have a part in this to help. We know that this is a great

step in faith and we are believing God to meet the need. So, please pray and be led by the Lord in what to give.

In the Love of God,

VOICE OF THE WORD MINISTRY

Evangelist Johnny and Sadie Ward and family[19]

Were this effort to prove successful, however, it would result in more travel for the Wards, with whatever family disruptions that might bring. "I worry about Wesley," Johnny says. "Next year he'll be in kindergarten, and the teachers won't like him missing a lot of school. We've thought about applying for one of those situations in which you can teach your child at home. We're looking at that."

Johnny's current role model apparently is the young evangelist Randy Smith of Millers Creek, North Carolina, a preacher he mentions frequently. Sadie gave me a copy of a "Randy Smith Revivals Newsletter," a legal-size document printed on both sides promoting this evangelist's ministry. On the front there is a picture of Smith, one hand above his head, the other holding a microphone, head tilted toward the heavens, obviously preaching with great exuberance. This side also contains an evangelical message from Smith and a few paragraphs written by Johnny Ward, suggesting that the two have worked together. On the opposite side there is a copy of an unidentified newspaper article about Smith, with photographs showing his tent, his wife and two pre-school-aged children, scenes of arm-waving worshipers at his revivals, and the converted school bus in which he transports the necessities of his on-the-road ministry and in which the family lives during revival-site stays.[20]

When I looked at the photograph of those children in that converted school bus, I asked Johnny Ward what kind of a future he wanted for Wesley and Melissa. "God told me," he said, "I would produce a prophet. So I'm hoping Wesley [*no mention of Melissa*] will receive such a calling. I'm not against education," he added. "I believe you should learn as much as you can. Still, God controls our lives."

Sadie Ward believes that God directs and protects the whole family. In a tract she distributes at revivals and other religious meetings she tells the following story. I have preserved the original emphases, grammar, spelling, and punctuation.

I lived in the clutches of Satan for 30 years, bound by sin, with no destination, my life a total wreck. It seemed my nerves could take *no* more. I had wanted children desperately. I had six miscarriages, and one girl and one boy that were born into this world only lived for a few days. I just felt that I could not go on.

My husband and I knew that the Lord was dealing with us, calling us to come into the fold, but I guess you could say we were running from the call. The devil was saying just end it all; life isn't worth living.

We were in a bad car accident, with the other driver drinking and in fault. I was driving our vehicle. The patrolman that came to the scene said it was one of the worst wrecks he had ever witnessed. The driver had hit us and on impact he was doing at least 95 miles per hour. My husband and I were both hospitalized. He was released after 3 days under strict doctors' care. But I, Sadie Ward, laid in the intensive care unit, not knowing anything . . . unconscious . . for a period of two weeks. Family came and did not know me from the apperance of my body. I had four operations on my face, and two on my stomach. But praise be to the Lord, I had a praying mother who believed in annointing with oil and laying hands on the sick and they would recover. I stayed a total of forty-one days. When I was sent home, my face was wired together; I could not eat anything and I was drinking water from a straw. My weight was 75 pounds. I was on pain medications 4 times a day and nerve medication 3 times a day.

The automobile accident was in October 1979. Months later I heard the most wonderful voice I had ever heard, and I knew it was the Lord! He was calling, calling, calling again, and I knew deep . . . deep within me that I had to make the decision . . . *heaven or hell.* I knew He had given me another chance.

I borrowed a dress and headed for the National Guard Armory in North Wilkesboro to hear Brother R. A. West. And there on that concrete floor, my husband and I gave our hearts *to Jesus.* My husband was a chronic alcoholic. We both received deliverance that night . . . *total deliverance.* We both received the Holy Ghost a few months later. And we have never been the same since December 16, 1982.[21]

Sister Sadie Ward's narrative communicates well the beliefs and values that undergird "The Voice of the Word" broadcast. Johnny Ward is determined to make his plans (which he calls "God's plans") for a career in traveling evangelism work. He visits tent meetings and studies the equipment involved, feeling certain that "Providence" guides him as he dreams. A radio ministry, he believes, is a vital part of the package, a way in which he can be heard by members of small churches at which he might preach a revival. In addition, he slowly establishes a network with other independent Holiness-Pentecostal evangelists, believing they can help each other in their respective missions. I accompanied the Ward family during one of these efforts at evangelistic networking, and what follows is my account of that venture.

Revival
A Trip to Drexel, North Carolina

As mentioned at the close of chapter 1, in each of these case studies I not only examine the respective broadcast, but I also seek to understand the airwaves-of-Zion individual or group in a setting or settings outside the radio studio. On January 4, 1991, a Friday, I spent the afternoon at Brother Johnny Ward's house in Triplett, interviewing the six individuals discussed above. That evening I traveled with this group to Drexel, Burke County, North Carolina, where Dewey was to preach a revival at a small independent Pentecostal fellowship, the Church of Miracles, Signs, and Wonders, All in Jesus's Name, pastored by Sister Kathy Benfield.

Leaving Simmons' Hollow, we took a back-country route out of Watauga County, North Carolina, driving southwest along a gravel road that follows Elk Creek until reaching Buffalo Cove. The road then moves through this narrow valley toward an intersection with Highway 321 just north of Lenoir in Caldwell County. From that point it was approximately a forty-minute trip to Drexel, Burke County, North Carolina.

The revival service had been scheduled to begin at 7:00 P.M.; but we were fifteen minutes late when we parked in front of the church, a forty- or fifty-year-old wood-frame structure consisting of just one room—the sanctuary—walled with tongue-and-groove planking that angled up into the slightly arched ceiling.

This inside meeting space reminded me of the multitude of starkly simple church interiors found throughout the rural South and Appalachia. The room was painted a light chalky gray; it was sparsely decorated, even severe in its Puritanical plainness; its austere illumination was provided by four bare light bulbs suspended from the ceiling; and it was heated—in winter—by a free-standing oil stove vented through the right wall. Two relatively small framed prints did break the disciplined unpretentiousness of the space, adorning the wall behind the pulpit with the only color I noticed—one of the Last Supper and the second of the face of Christ. Brother Johnny Ward might call this structure a "no-nonsense house of God, a Praise-the-Lord church."

The pulpit sat on a platform raised some seven or eight inches above the main floor, sharing that space with the church's sound system—a large amplifier, a tape player, several speakers, three microphones, and a number of electric musical instruments. Dewey Ward brought with him a second amplifier to be used with the instruments he and Daniel play.

There was no altar rail to separate the preaching area from the "pit" (space between the pulpit and the pews) and the audience area, suggesting by this absence both a structural and a spiritual unification of these three regions. As will be demonstrated by this account, Holiness-Pentecostal worship involves congregations in a fully participatory fashion, even to the degree that the center of activity occasionally is pulled to the pit or to the pews. Even when this is not the case, all formal divisions between the preaching space and the auditing space are significantly reduced.

An upright piano sat just off the platform to the left, with a microphone positioned to allow the pianist to join her voice to the amplified sounds. Later in the service the top of the piano was raised and a mike placed there. All seemed to be in readiness for a service filled with electronically intensified expressions of celebratory religiosity.

When we arrived, Sister Kathy Benfield had already assumed a position behind the pulpit, speaking passionately—microphone in hand—to a group of only eight worshipers. As exemplified by the presence of all the microphones and speakers, this church fits well that Pentecostal/Holiness/Freewill-Baptist pattern that calls for heavily amplified sound during preaching, singing, and testifying, creating an atmospheric environment in which volume equates with enthusiasm, happiness, commitment, and presence of the Spirit, and in which every effort is made to reach the limits of emotional expression. Indeed this amplified-sound factor will become a matter of discussion in all four of the case studies of this volume.

Saying that she did not intend to preach that evening—that Brother Dewey would play that role—Sister Benfield continued to hold the pulpit for another fifteen minutes, testifying with explosive acclamations of joy that energized her own kinetic expressions, while at the same time generating shouts and arm-waving jumps and hand clapping in others, even though the audience was small. It became immediately obvious that this gathering of worshipers had come to this church primed to make the fullest demonstration of their religious fervor.

During the most exuberant peaks in her moments of testimony, Sister Kathy Benfield fell into glossolalia, suddenly uttering those

Sister Kathy Benfield, *far center,* leading her congregation in singing.

fast-paced strings of (for me) unintelligible sounds that are identified in the Holiness-Pentecostal tradition as "tongues," and in the process exemplifying her acceptance of Mark 16:17: "In my name . . . they shall speak with new tongues." Practitioners of Holiness-Pentecostal worship also point to 1 Corinthians 12:10, 13:1, 14:2, and 14:10, and Acts 2:4–15 in support of glossolalia. Sister Kathy Benfield's distinctive contribution to these moments of "tongues" was that she accentuated these utterances with a series of rapid side-to-side shakes of her head, keeping her eyes tightly closed as she executed these movements.

A full-figured, dark-haired, strong-voiced woman in her mid- to late forties, Sister Kathy engaged in unbridled physical demonstrations of her emotions, utilizing the entire platform for her movements, and occasionally stepping off that structure and moving closer to the pews, where one woman always responded to those proxemic thrusts with her own standing, jumping, hand-waving dynamics. Indeed, this one audience member was not the only such respondent. Several outbursts of applause and shouting occurred even in this early part of the service, and emotions were peaking long before Brother Dewey Ward took his own position at the pulpit.

As mentioned above, the smallness of this audience did not restrain behaviors in any way. Indeed, the impression I received was that every worshiper felt obligated to be as demonstrative as possible, returning spiritedness with spiritedness, ardor with ardor. One authority on Pentecostal practices has said simply that "worship services tend to be evaluated on the basis of the amount of outward action."[22]

Appalachian Holiness-Pentecostal preaching is not prone to be as rhythmical and poetic as the sermonic modes I described in *Giving Glory to God in Appalachia* and *The Old Regular Baptists of Central Appalachia*.[23] Many of these preachers do break their exhortations into the short, punchy, "Haah"-accentuated lines I have spent so much time describing elsewhere, but the more-cadenced or melodic sounds found in the preaching of certain Appalachian "Old Time" Baptist subgroups are generally not present. Instead, Appalachian Holiness-Pentecostal delivery is dominated by a high-volume, rapidly paced, but prosaic and explosively emotional, exhortation that features a constant repetition of certain end-of-sentence exclamations ("Praise God!" "Amen!" "Hallelujah!" "Glory!" "Oh, Lord Jesus!" "Save us, Lord!" "Praise His name!").[24]

Content places primary emphasis upon the personal joys of salvation, with these joys frequently set in sharp relief against reminders of the consequences of nonredemption. Precise doctrine is rarely examined in detail, as preachers satisfy themselves with proclamations of the "Holy Ghost anointed" powers of Spirit. Emphasis is placed on the elicitation of auditor responses, with the preacher often calling for congregational affirmations of what he or she has said: "Let me hear someone say 'Amen!' to that." "Somebody needs to applaud that." "If you agree let me hear a 'Hallelujah!'" "Everybody who loves Jesus I want to hear you shout it right now. 'I love Jesus!'"

Finally, there is the hyperactive nature of the physical delivery-constantly in motion; accentuated by forceful surges of movements that constitute jumps, quick sprints, and bounding hops; vigorous thrusts of arms and head; and jerking tremors that possess the entire upper torso; all of this usually accompanied by verbal expressions of great joy. Indeed, there appear to be no limits to these demonstrations of the speaker's exuberant spirituality.

These unrestrained physical movements on the part of the preacher or testifier also, as suggested above, engender comparable behaviors within the audience; and practitioners of independent Holiness-Pentecostal worship experience cathartic highs seldom realized in mainline religious expression. For good or for bad, these individuals yield completely to the passions of their faith. They break into applause, shout without saying anything in particular, cry out "Amen" or "Praise God," speak in tongues, "dance in the Spirit," "swoon in the Spirit" (also known as "slain in the Spirit"), wave their hands above their heads, and just generally exult in the joyous emotions of testimony and preaching.[25] And if they do not respond in some of these fashions, the preacher, as I indicated earlier, is apt to ask them to do so: "Come on, Church! Let's hear some shouting for Jesus. A quiet church is a dead church, a church without Jesus."

After her own testimony Sister Kathy called on individuals and groups to sing or testify, including the six persons with whom I had come to this event. With the exception of two female members of the audience, the children (Wesley and Melissa), and me, all worshipers (now seventeen in all) either sang, played a musical instrument, testified, or all three. The principle suggested by these individual involvements was that each person assumed a responsibility for keeping the service going: Each was there not to be performed for or to be spoken to, but to make an individual contribution to this striving for "anointment." Those contributions become essential, because independent Holiness-Pentecostal services—at least in Appalachia—tend to have no preset form and simply evolve out of the impromptu contributions of the respective worshipers.[26]

After about forty-five minutes of general singing and testifying, Brother Dewey Ward began his preaching. One of the microphones had a twenty- to twenty-five-foot cord leading back to the amplifier, enabling a preacher or testifier to roam freely across the entire front of the meeting space and to step into the pit or into the audience area. Brother Dewey immediately took possession of that mike, moving quickly into the exercise of a physical style that never kept him in one place for longer than a few seconds. After only a minute or two of his preaching it became clear that he would have trouble keeping that long microphone cord from becoming entangled with various objects on the platform, including the pulpit; so Johnny Ward went to his aid, flipping the line back and forth across objects as his brother made his quick surges from one side of the meeting space to the other.

Like that of Sister Kathy, Brother Dewey Ward's platform style was very kinetic, employing arms, legs, shoulders, head, hands, feet, and complete torso in free-flowing, unrestrained sets of dynamic expression. Occasionally he would stand in one place and simply throw his arms, head, and shoulders back, bending his legs slightly to keep his balance, and send forth some soaring shout of exultation, praising his God for his wonders and mercy. At other times he would also hold in one spot, but pull his body inward—eyes closed, face grimacing, chin tucked tightly against neck, shoulders pulled forward and inward, hands fistlike to each side of his head, body crouched low and rigidly—to denounce the horrors of Satan's guile. Such individual moments of passion were linked by furious surges across the performance space—transitional movements usually accentuated by stamping strides and sudden leaps or sprints and by impassioned verbal statements. Always in the background there was Johnny Ward, manipulating the mike cord to avoid possible snares, the two brothers working in an interesting kind of kinetic union, connected in space by that thin line of wire and polymer.

Audience members responded in kind, seldom seated for very long, and often moving into tight standing clusters, elbows out and hands slightly above their heads, swaying or shaking. Some members appeared to sink into their own personal worlds of rapture, eyes closed, faces captured by frozen expressions of joy, excitement, or tranquillity—sometimes suggesting paroxysm and calm in a marbleized mixture.

In the midst of all this audience reaction, I began to notice one woman sitting on the front pew to the far left. What drew my attention to her was the contrast between her behavior and that of other worshipers occupying the front pew. This area had become the focus of responses to Brother Dewey's preaching, with Sister Kathy and another woman seeming to lead audience feedback. The woman in question, however, had remained seated throughout the sermon, even during the most boisterous moments of the evening. Occasionally I did see her lean forward and tense her shoulders to such a degree her muscles trembled, but she held to her seat when all the worshipers around her were standing, clapping, shouting, and crying.

Experience in observing such events led me to suspect that the woman eventually would go forward, yielding to the emotions of the evening, collapsing in the pit area, crying uncontrollably, and declaring—between sobs—both her sinfulness and her acceptance of a "saving grace." That was precisely what happened, but not until Dewey Ward closed his sermon and began an impassioned altar call.

In fact, it took considerable urging on Brother Dewey's part, directed to the worshipers in general, before this woman responded. Employing several traditional techniques of evangelistic altar calls, Ward first went through three verses of an invitational hymn, constantly entreating the "unsaved" to come forward. Finally he asked all audience members to bow their heads and shut their eyes. With this condition of anonymity established, he then pleaded for anyone troubled over the state of her or his soul to raise a hand. The piano played softly in the background, accentuating the emotionality of this appeal, and one lone male voice in the congregation kept repeating, "Won't you go? Won't you go? Jesus wants you. Won't you go?"

Apparently the woman under study did lift a hand, or someone did, because Ward then pushed even harder to encourage some individual forward: "You held up your hand, admitting your need for Jesus. Now you should come to his throne of grace. Come to Him. Come to Jesus." The piano seemed to make the same plea, increasing the intensity of the moment. Soft weeping could be heard coming from one woman, accentuated at times by more spasmodic cries. One lone shout was heard that faded into a sob.

It took more urging, but eventually the woman in question did step forward—more like springing forward—and a shouted "Praise God" from Dewey signaled to the eyes-shut audience that some breakthrough to "salvation" had been made. Suddenly all members of the congregation were alive, heads-up, and standing, captured by a new intensity in the meeting. Passionately felt emotions surged through the actions of this moment, an onrush of elation that billowed and broke upon all these believers. "Amen! Hallelujah! Praise God!" repeated Brother Dewey.

Three members of the congregation were immediately around the kneeling woman, hands on her head, back, and shoulders, crouching low and talking to her, drawing from her those even more forceful cries of grief, joy, or a mixture thereof, occasionally embracing her, but never losing that physical contact with her. After a few moments on her knees, the woman stood, helped to her feet by the three supporting revelers, who in the process seemed to draw even closer to her. Now she was positioned before Brother Dewey, who had found a place on the forward lip of the raised platform.

Though standing, the woman was still stooped and shaking, her entire body racked with emotion, and her face buried in her hands, not saying anything that I could hear. Still her mere presence at that spot cried out a message to those around her. She had come to the place where "sinners" come—"blackened by heaven-denying transgressions, the sins of natural man, the sins of Adam," but abject and yielding—to be forgiven, to be cleansed, to be redeemed. Now she waited, her posture tensed into trembling.

Almost immediately she was surrounded by several additional members of the congregation, each making his or her own impassioned response to this event—shouting, praising, crying with joy. For about ten minutes this group reveled in the circumstances of this moment, everyone hugging the woman and each other, individuals taking their turn at having hands laid on them by Brother Dewey.

It was an immensely high moment for all, transcending almost all events in church fellowship. A new Sister had joined the "Saints," reaffirming by her action the validity of this energy and passion—all the singing, testifying, crying, and shouting—all the personal investments in prayer and exhortation. This was the moment when a "soul was saved" and "Jesus took command." It needed to be felt and celebrated by all.

The woman's mother, one of the first three who had been around her, renewed her own crying and testifying, rejoicing that "the Lord had returned her daughter to her"—in two different ways, she said: a "runaway child" had come home and was now "seeking salvation." The father was also there, in the circle of rejoicers, nervous and crying, but not as demonstrative as his wife and appearing somewhat uncomfortable with the moment, not nearly as sure of his role as his wife seemed to be of hers. Earlier in the service he had demonstrated

The woman, her hands raised, in front of Brother Dewey Ward, Drexel, North
Carolina.

his value to the congregation by taking care of all the sound equipment, adjusting the readings on the amplifier whenever the speakers
squawked. Unlike his wife, however, he had not played a role in the
emotional displays, suggesting that such demonstrations were not
his forte.

From the mother's tearful testimony we learned that the daughter
had been lost to the family for several years, her mother and father
not knowing where she was or if she still lived. Then a phone call
came, only days before this present event. The daughter wanted to
come home. Would she be welcome? "Yes!" cried the mother—at the
time of the telephone conversation and again before this gathering.

"When she came to the house today, she was like a street person, a
dirty hippie," the mother cried. "She had to scrub herself clean before
she could come with us tonight. Praise God! Oh, Jesus! Praise God! I
love you, Lord! This is the happiest moment of my life!" For a moment
two other women supported the shouting mother as she now gave full
vent to her own joy.

Concurrent with these actions, a final burst of passion developed
as several individuals in the group "laid hands on" each other in the
traditional posture of healing or blessing. This collective energy and
ardent expression of exaltation pulled all but three persons to that
spot in front of Brother Dewey Ward, there to be emotionally charged.

There was no official closing of the service. Everyone seemed to
sense that this was where it ended. Nothing more could be said or
done that would top the emotions of this "coming forward." In a very
real sense the congregation was drained, a full catharsis having occurred. Eventually the young woman sat with her mother on the front
pew, in quiet communion, an occasional sob still being heard, but
muffled now. The remaining members of the congregation milled
about, reluctant to let this moment end, but too spent to do otherwise.

It took a few minutes for the time to be right, but eventually I approached the young woman. I will not give the woman's name because
of the sensitivities of this account; however, I will observe that she
was young, probably in her mid-twenties, attractive, but showing
some of the wear and tear of a not too gentle or placid life. She knew
who I was. Sister Kathy Benfield had, much earlier, acknowledged
my presence at the service, saying that she understood from Johnny
Ward that I was a writer "of some sort."

During the subsequent interview the daughter said simply that
she had been traveling for several years. She mentioned California,

Colorado, and "other states." What had she done; how had she survived, I asked.

"A lot of things of no importance," she answered, then pausing for a long moment. "For a while I was on the streets," she finally added, her voice dropping. "I ain't proud of that, but it's what I done to live."

"What does tonight mean to you?" I continued.

"Freedom," was her response, starting to say something more, then stopping.

"From what," I probed.

"From guilt, from shame, from separation from my folks," she choked, beginning to cry again.[27]

I caught a glimpse of the mother and thought I saw concern in her face; therefore, I stopped this interview and thanked the woman for her forthrightness. Nothing was to be gained from additional questions that troubled the family. Besides, the woman was crying again now, her face buried in her hands. For at least ten minutes she continued to sit in her pew, everyone now allowing her to cry alone, even though her mother sat close and watched.

Our Trip Back to Boone
The "Loose Wheel" People

We seven adults and two children had traveled from Simmons' Hollow to Drexel in three cars, an arrangement necessitated in part because Dewey and Cindy Ward, along with Daniel and Stephanie Riddle, needed to return that night to Elizabethton, Tennessee, and did not need to retrace the route to Johnny Ward's house. The same thing was true in my case, so I asked Dewey and Cindy to ride to Boone with me, with Daniel and Stephanie following in Dewey's car, the line of travel these four would need to take anyway.

When he climbed into my car Dewey Ward was still experiencing the emotional and psychological elation he had reached at the end of the service. "Praise the Lord!" he said. "Praise the Lord! That girl coming to the altar made my day, praise God." He spoke with such enthusiasm I had no difficulty understanding that such events meant much to him: he had been an instrument in a process he considered of great importance—monumental importance. "Jesus saves," Dewey Ward had frequently proclaimed in his sermon, but how would this evangelist have characterized his own role? I did not ask him this question, but I wish I had.

After Brother Dewey's exhilaration gradually subsided, our conversations switched to other questions about his work and about Pentecostalism. One topic we broached concerned Pentecostals and individual church commitment. I told him that one criticism I had heard of Pentecostals was that their passions for religious expression were strong but that their involvements with individual churches (as formal members of particular fellowships) were weak, all emotion and no hard work to build the church as an institution.

"A lot of folks think Pentecostals are loose wheels," he said.

"Because you roll so freely from one church to another?" I asked.

"That's about it," he added. "They think we ought to join one particular church and commit ourselves to that one group of people. Instead, some of us wander around a lot, attending a lot of different churches. Joining a particular church is not all that important to us. We go where Jesus calls."

"The people as well as the preachers?" I asked.

"The people as well as the preachers," he answered. "We get lots of people who don't fit well in the other denominations. They're not accepted. A lot of poor people. The folks who get emotional, who like to cry and shout. Who like to testify. They find a place in our services. Still, they don't stay put in any one church. The little church I've started in Elizabethton doesn't even have members: it has people who come to praise God. We don't bother to write names down anywhere."

As I listened I recalled a conversation I had with a Kentucky Old Regular Baptist, a subdenomination that places great importance upon individual church membership and commitment to a specific fellowship, a fellowship that formally admits to membership and just as formally "excludes" from membership, with all the controls over personal behaviors thus involved.[28] This was a discussion during which the man examined some of the other religious groups of which he was aware, and I must add that this Old Regular possessed an especially strong disdain for the Holiness-Pentecostal groups of his area.

"They get together on Friday and Saturday nights and bang on their musical instruments, and even get up and dance, all of that looking like entertainment to me. Then they don't build and maintain long-lasting fellowships," he charged. "They start a church in some abandoned store, and it goes for a while. But soon you'll see that place has closed and everybody has gone some place else."[29]

The "loose wheel" metaphor, as applied to Appalachian Pentecostals, was new to me, but the charge that these people were not firmly anchored was not new, as exemplified by the remarks of the Old Regular Baptist, a conversation that I did not mention to Brother Dewey. What I heard Dewey Ward saying to me that evening was that there is a noticeable body of Holiness-Pentecostal worshipers who not only do not feel comfortable in mainline churches but who also do not find permanent homes even in Holiness-Pentecostal congregations. Call them "loose wheels" (Ward's term), or use Robert Mapes Anderson's language, "the disinherited," and you appear to be describing the same characteristic.

Anderson even found a pattern of "shifting allegiances" in the religious lives of forty-five leaders of the Pentecostal movement during its earliest years, 1909 through the 1920s. He wrote of their "mobility . . . and marginality: spatial, occupational, and religious. Wandering from place to place, from job to job, from church to church."[30]

Furthermore, as suggested by the Old Regular's comment, one factor contributing to this image of instability, wandering uprootedness, and high turnover in fellowships arises out of the multitude of small denominationally independent churches many Pentecostals attend, churches that frequently do not have "Pentecostal" or "Holiness" anywhere in their names, perhaps only the words "Full Gospel," and perhaps not even that revealing term. The sign in front of Sister Kathy Benfield's church, for example, gives no indication at all of its denominational identification: "The Church of Miracles, Signs, and Wonders, All in Jesus's Name." The same is true of Brother Dewey Ward's church in Elizabethton, Tennessee: "Jesus Faith Center." It is also true for the church of Brother James R. Kittinger, to be examined in the next chapter.

This confusion regarding church names is further accentuated by the "bewildering array of Pentecostal sects" that David Edwin Harrell, Jr., mentions, an array that is categorized by Anderson.[31] When my Old Regular acquaintance spoke of the apparent inconstancy of the Holiness-Pentecostal people he knew, he may have been voicing his frustration in determining exactly what was the faith of these people, given their attendance at churches with such widely differing titles.

Finally, the Old Regular Baptist's charge that these churches often spring up suddenly and then disappear just as quickly also deserves some consideration. My conversation with Brother Dewey revealed that he was not at all certain he would be able to keep his Elizabethton, Tennessee, Jesus Faith Center afloat. Brother Dewey does not own the facility, and the woman who does could withdraw from the agreement she has with Dewey Ward should she become disen-

chanted with his ministry. In addition, there is no outside adminis-
trative structure to help this small fellowship: the church is a totally
autonomous operation, not dependent upon or supported by a confer-
ence or an association. The same is true relative to Sister Benfield's
church, and that small fellowship is constantly struggling to stay viable. According to Benfield, the steady attendants dropped to only
three at one point in 1990. Would she be able to keep the doors open? "I
pray that I can," she said. Still, she did not seem inordinately confident. That body of believers I saw on Friday, January 4, 1991, did not
appear numerous enough to sustain Sister Kathy Benfield's ministry.
In addition, would they soon grow tired of her preaching and wander
somewhere else?

Holiness-Pentecostal followers, however, would quickly argue that
a church building is not needed, that their worship can take place
anywhere, particularly in private homes. We will see indication of
that practice in chapter 5 when I discuss Sister Brenda Blankenship.
In addition, remember that Brother Johnny Ward began his involvement with this faith by attending services at his sister's house in
Mount Pleasant, North Carolina. Holiness-Pentecostal practitioners
tend not to place as high a value on a church facility as do fellowships
of mainline faiths. So those buildings can come and go and the Spirit
still remain.

Back to "The Voice of the Word" Broadcast

Brothers Johnny and Dewey Ward carry into airwaves-of-Zion radio
stations the same passionate and tempestuous evangelism that they
practice in the small independent Holiness-Pentecostal churches of
Appalachia, turning studios into arenas for "foot-stomping Holy
Ghost" advocacy. The result is religious programming that pounds and
pulsates, frequently lacking clarity and logic, but never lacking zeal
and animation. It is the type of broadcasting that the fastidious would
avoid with the quick turn of a dial, the push of a button. Nevertheless, it
has its following and its place in America's folk culture, particularly in
Appalachia.

Station owners and program managers frequently find themselves
in the awkward situation of disliking these more boisterous airwaves-
of-Zion products but having to air the programs simply because there
is still a market for the manner and the messages of preaching in the
style of Johnny Ward's "latter rain" gospel. At WMCT, particularly, I
sensed Fran Atkinson's fear that my attention to this specific program would made it "representative" of the station at large. Perhaps

that situation, alone, indicates a limited lifetime for this genre of religious programming.

As I have suggested by this treatment of Brother Johnny Ward, no adequate ethnographic examination of one of these programs can begin and end with the broadcast itself. The activities of these airwaves-of-Zion individuals and groups extend outward from these radio studios into much larger areas of religious involvement. An examination of the Ward family and the "Voice of the Word" broadcast must consider that larger socioreligious setting. In turn, that broader framework reveals a complex reality that is not easily assessed. How, as examples, does one evaluate the "coming to Jesus" episode that developed in that small church in Drexel, North Carolina, or the effects of this Ward family mission on the two children involved?

I will not attempt any personal judgment of Brother Johnny Ward and the various individuals with whom he works. I will leave that task to the reader. Ward himself is very tolerant of religious diversity, both in doctrine and practice, repeating over and over again that his role is not to divide. Both Johnny Ward and Dewey Ward, however, are troubled by the "Holy Roller" negative images that prevail among mainliners. Nevertheless, like the southern Indiana Pentecostals examined by Lawless, they expect these negatives to be countered by a fresh outpouring of "Latter Rain," a new Pentecost that will reveal the validity of their "Holy Ghost possessed" practices.[32] They point to James 5:7–8:

> Be patient therefore, brethren, unto the coming of the Lord. Behold, the husbandman waiteth for the precious fruit of the earth, and hath long patience for it, until he receive the early and latter rain.
> Be ye also patient: stablish your hearts: for the coming of the Lord draweth nigh.

Like most Pentecostals, they argue that the "early rain" was the first Pentecost and the "latter rain" is the one yet to come—or is at this moment coming—when the prophecy found in Joel 2:28 will be fulfilled:[33] "And it shall come to pass in the last days that I will pour out my Spirit on all flesh." So, in the words of Joel, they "blow the trumpet in Zion," over the airwaves of Zion.

A tent and a trumpet: Those are the needs of Brother Johnny Ward. He has his trumpet, in the form of his "Voice of the Word" broadcast, muted though it might be; and now he hopes for a tent, and perhaps a converted school bus in which to transport his family and all the equipage of his evangelistic mission.

Rex and Eleanor Parker and "Songs of Salvation"

We've been laboring so hard, and Lord we're awful tired
Just trying to get some poor sinner saved.
Won't you let Him come in; He'll free you from you sin,
Then you'll be happy everyday.

Rex and Eleanor Parker[1]

At the time of this writing, Rex and Eleanor Parker live on Bent Mountain Road, near the small community of Lerona, Mercer County, West Virginia, on twenty-seven acres of Appalachian pasture and woodland they call "the farm." Their home is a well-lived-in structure, neither squalid and ramshackle, nor idyllically rustic and tidy. It is forty to fifty years old, needs paint and repair in spots, and is secured by multiple chains and padlocks on every outside door, the latter being Rex's response to a burglary that cost him several prized hunting rifles and shotguns. "I know who did it. Can't prove it," he said, "but I don't intend for it to happen again."[2]

Surrounded by a number of aging cars in various stages of dismantlement—objects of Rex's avocation, automotive bodywork—the outside of this home fits well those junkyard aesthetics that characterize some areas of central Appalachia, particularly the coalfield regions of eastern Kentucky and southern West Virginia. Inside, however, the environment possesses a more traditional charm and picturesqueness, with walls and shelves covered with the artifacts of two significant influences on the lives of this couple, religion and music. The first of these influences is represented by numerous plastic, plaster-of-paris, or black velvet religious plaques, hangings, and inexpensive bric-a-brac; and the latter is exemplified by photographs of the Parkers and their children taken during the years they played bluegrass music professionally and by one bedroom literally filled with stringed instruments, most of which Rex crafted himself.

Rex (his adopted show name) is, at the time of this writing, seventy, and Eleanor is sixty-nine. Married fifty years and partnered performers for the same amount of time, they have been what Eleanor calls "childhood, adulthood, and foreverhood sweethearts."[3] They have also been, for the past thirty-three years (since they "found salvation"), impassioned practitioners of a "come-to-Jesus" evangelism—nondenominational, fundamentalist, deeply joyous, Appalachian, and fervently expressed through "Songs of Salvation," the title not only of their hour-long Sunday show over WAEY-AM in Princeton, West Virginia, but also of their thirty-minute Saturday program over WJLS-AM, Beckley. With their ages showing, Rex and Eleanor Parker occasionally do look "awful tired," but they give no signs of wanting to conclude their airwaves-of-Zion mission, their "laboring so hard."

WAEY-AM/FM, the "Country Way"

I first met the Parkers when I visited WAEY on April 16, 1989. At that time I was sampling the Sunday morning and afternoon fare of Appalachian AM stations to identify live religious broadcasts I would later examine in detail. A telephone call to the station's general manager had produced an invitation to visit the WAEY facilities and a suggestion that I arrive by eight o'clock on Sunday morning if I wanted to catch "Songs of Salvation," characterized by that individual as the most interesting of the station's live religious broadcasts.

At the time I write this volume the management of WAEY promotes their operation as the "No. 1 Station in Princeton, Bluefield, & Mercer County, West Virginia." This AM/FM affiliate simulcasts its programming on both bands, except on Sunday when the two programs split to their respective frequencies, leaving these Sunday AM programs with a much smaller audience than they would otherwise have enjoyed.

With twenty-four hours a day of broadcasting, WAEY-AM/FM follows an exclusively country-music format hailed as "Your Country Way," reaching—via its FM transmissions—seven counties in West Virginia (Mercer, McDowell, Raleigh, Summers, Monroe, Greenbrier, and Wyoming) and six counties in Virginia (Tazewell, Wythe, Bland, Giles, Montgomery, and Carroll). Currently, WAEY's weekday schedule is built around the following shows, described here in the words of the promotional materials the station distributes to prospective advertisers:

> 6:00–10:30 A.M. THE BUDDY ANDREWS SHOW—The Country Music Breakfast Club, featuring the top chart Country sounds, selected country classics, homespun humor and the latest news of the Country Music Industry and artists from Nashville. . . .
>
> 10:30–3:00 P.M. THE DOTTI O'DALEY SHOW—Country Music continues on the "O'Daley Show" with the top country sounds, extra emphasis on the programming of artists and news of the country artists and their families. . . .
>
> 3:00–7:00 P.M. THE LEE DAUGHERTY SHOW—"Mrs. Daugherty's Little Boy, Lee" rolls through afternoon drive time with a fast paced presentation for the Southern West Virginia area, and current sports information for the up-coming evening. . . .
>
> 7:00–12:00 M [midnight] THE AL HARMAN SHOW—"Big Al," as he is known to his listeners, opens the request time from 7 P.M. until midnight with an instant request format. Al features the top 100 country sounds, and a complete library of Country Classics. His interesting comments directed toward, and about, his request line callers make them feel a part of the show. Complete coverage of all sporting events, local and statewide, are a featured part of "The Big Al Show."
>
> 12 M–6:00 A.M. THE NATHAN HUNT SHOW—Nathan continues on thru the wee hours of the morning playing the best of country and listener requests. News and weather on the hour plus local school closing announcements and road conditions as information is received.[4]

WAEY-AM's Sunday programming features gospel music and a series of live groups, the most interesting of which are the following: "Songs of Salvation" (8:00–9:00 A.M.), soon to be examined; "The Jesus Name Trio and the Conley Sisters" (12:00–1:00 P.M.), an impassioned hour of preaching, singing, testimonials, and prayer, led by Sister Edna Conley, a Pentecostal evangelist of Bland, Virginia, with the assistance of five singer/musicians—Becky Conley, Shirley Conley, Linda French, Cathy Stevenson, and John Stevenson; and "The Gospel Light Broadcast," the program discussed in chapter 1 as an example of the practice of acknowledging countless infirm listeners. One additional broadcast rounded out the live programming on that Sunday in April 1989, when I first visited WAEY: "The Rev. David Barnette Program" (2:00–3:00 P.M.), produced by the named evangelist, pastor of the Free Pentecostal Church of God fellowship in Bluefield, West Virginia.

"Songs of Salvation" features Rex and Eleanor Parker, along with

J. Howard Kress, an acoustical guitarist who has played with Rex off and on since 1937. During these radio broadcasts, Rex usually plays either guitar, mandolin, or banjo, while Eleanor moderates, picks, or strums a second guitar, presents the show's advertisements, and sings solo or in harmony with one or both of the men. In addition to his own guitar accompaniments, Kress engages in humorous repartee with Eleanor between songs and during the long commercials Eleanor delivers.

These commercials make "Songs of Salvation" distinct from other airwaves-of-Zion productions I examine in this volume, for the show is sponsored by business establishments whose products and services are promoted by Eleanor. Furthermore, it is the homespun and 1940s quality of these commercials that provides much of the charm of the broadcast. Indeed, older listeners who hear "Songs of Salvation" for the first time will be taken back in memory to the broadcast styles and techniques of the World War II era.

Although neither Eleanor nor Rex write music, they have composed most of their songs, imbuing them with either an intense religiosity or an almost turn-of-the-century tone of melodrama. Occasionally the themes of these compositions treat a crippled child, a drunken father, an aging mother, or a destitute soul, with all problems being solvable—or bearable—through an alignment with God:

> If you're a slave of sin, fall upon your knees,
> Come and ask the Savior and He'll set you free.
> Don't go on in sin and meet that dreadful end,
> You're welcome to the Kingdom, but there's only one way in.[5]
>
> The Parkers

In one respect, these aging musicians represent a rapidly diminishing breed of central Appalachian radio performers (country-western, "hillbilly," then bluegrass) who flourished from the 1930s through the early 1950s, broadcasting in many cases to mine-worker audiences. These performers hawked everything the nation's business sector had to market—flour, baking soda, illustrated Bibles, home furnishings, new and used automobiles, life and health insurance, songbooks and records, farm equipment and fertilizers, produce, animal feeds, gasoline and motor oils, mortuary services, soft drinks, coffee, and a host of patent medicines with names such as Hamlin's Wizard Oil, Bi-Tone Wonder Tonic, Black Draught Tonic, and Hadacol. In particular, Rex and Eleanor Parker represent a rich vein of country-and-western and bluegrass musical talent that developed around the southern West Virginia towns of Bluefield, Princeton, and Beckley during the 1930s and 1940s, performing on such stations as WHIS (Bluefield), WAEY (Princeton), and WJLS (Beckley).[6]

In another respect, however, Rex and Eleanor Parker are appropriate subjects for this examination of the airwaves of Zion, having left the bluegrass performing circuit and devoted themselves to a calling centered in church sings and religious broadcasting, the latter produced in that richly traditional, 1940s-sounding format and style. Through regular weekend programs on WAEY and WJLS-AM, Beckley, West Virginia, the Parkers send forth across the "radioland" of southern West Virginia and southwestern Virginia a warmly textured mixture of country-and-western/bluegrass gospel music, prayers for the sick or downhearted, humorous but inspirational anecdotes and maxims, and such commercial messages as Eleanor says she can personally endorse. Indeed, this couple once dropped a sponsor who started selling beer, wine, and lottery tickets, products the Parkers cannot bless. That establishment has since discontinued traffic in these objectionable items, and at the time this chapter is being composed Rex and Eleanor are negotiating new arrangements with the firm.[7]

Charles (Rex) Parker was educated only through the third grade, and is a self-taught musician who has mastered a great variety of stringed instruments (including guitar, banjo, fiddle, mandolin, bass fiddle, dulcimer, among others). He began his professional career in his late teens on the "Old Farm Hour" at WCHS in Charleston, West Virginia, before gaining additional broadcast experience at WJLS in Beckley, WHIS and WHIS-TV in Bluefield, and WOAY-TV in Oak Hill, performing at one time or another with such country music legends as Lloyd "Cowboy" Copas, T. Texas Tyler, Little Jimmy Dickens, Dixie Lee Williamson (who later obtained her national following under the name Molly O'Day), the Stanley Brothers (Carter and Ralph), and Bill Monroe.[8]

With a non-assertive and somewhat shy demeanor, but a perfectionist's attitude toward his art, Rex (known as Curly in his initial performing years) never was the leading personality in any of the groups with whom he performed, and when he teamed with Eleanor he let her emcee their shows and provide the comic repartee characteristic of bluegrass and other country-music stage acts and broadcasts. "She's always the talker," he says, and he lets that stand as his reason for not saying much during performances or interviews. A serious instrumentalist, he concentrates on his playing even to the point of looking rather dour while rapidly fingering mandolin or banjo tunes, suffering a look of pleasure only after the completion of a piece of music. "They told me I should smile," he says, but he seldom manages to do so until he sings (for humor or for inspiration), at which time he occasionally exhibits a grin which could be described as impish.

Eleanor, on the other hand, is always smiling, always chatty, always energetic, and always ready to throw out witty, lyrical, and occasionally rhyming phrases obviously pulled from a long-established repertoire of banter. "He ain't too fat, and he ain't too thin; he just pooches out where he ought'a pooch in," she proclaimed on the air about J. Howard Kress, responding to the latter's statement that he could not eat many pancakes any more.[9] "For the dealingest dealer that ever dealt a deal it's Harry Beal," she announces in her WJLS advertisement for Harry's Mobile Homes.[10] Concerning a WAEY sponsor, Mills' Market, she says,

> If you want it, name it.
> If they've got it, you'll get it.
> If they ain't, they'll get it.
> And if they can't, then you forget it.
> Because you probably don't need it nohow.[11]

During years of travel on the West Virginia, Virginia, Kentucky country-music circuit, she developed a comic spiel sprinkled with lyrical ruralisms that apparently delighted her 1940s audiences, and which even today entertain her aging but still unsophisticated listeners. Faithful followers of the Parkers' radio broadcasts hear certain catchy phrases or jokes over and over again: "Rex and I have the two sure signs of getting old. One is you forget to remember, and I'll declare I can't think what the other one is." "J. Howard, you're not real old. You just started out real young." "Someone once said to me, 'How in the world did you stand looking at that man [Rex] across the table for forty-nine years?' I said, 'The very same way he stood looking at me.'"[12]

If this repetitiousness bothers the WAEY and WJLS airwaves-of-Zion listeners, they do not tell Rex and Eleanor. Perhaps many of these fans relive, via these often repeated lines, those earlier and better years when Rex and Eleanor Parker were, as Eleanor now claims, "bywords" in southern West Virginia.[13]

Eleanor Parker (formerly Eleanor Neira) met Charles Parker in the mid-thirties, and the two became, according to Eleanor's frequent declarations, "childhood sweethearts." During an era in which thousands of immigrant labors were employed to work the Appalachian coalfields, her father had come to West Virginia from Barcelona, Spain. Eleanor is unable to name the exact year this happened, but apparently it was sometime in the second decade of the 1900s.

Neira left his wife and several children in Spain for six years while he mined West Virginia coal, saving money for an eventual reuniting of the family. Born in 1922—after that reuniting—Eleanor grew up in a family of nine singing siblings. The Neira children entertained themselves with Spanish songs taught by their mother, who accompanied her offspring with a tambourine.

Eleanor's mother never learned English, so all of the youngsters became bilingual. However, that cultural and linguistic factor did not prevent Eleanor from developing musical tastes and talents prevalent in the West Virginia mountains, although in later years she frequently added Spanish songs to the Rex and Eleanor repertoire. At the present there appears to be little or no dialectal trace of those early years of speaking another language. Indeed, the speech of Eleanor Parker falls very much within the West Virginia idiom, and one hears evidence of her Spanish speaking skills only when she discusses her parents' origins: Her pronunciations of Spain's cities and provinces demonstrate her tongue's control of Castilian sounds. Although her parents were from Barcelona, the family apparently did not speak Catalan.

At the close of her elementary schooling, family circumstances moved Eleanor from West Virginia to New York City, where she lived with two older sisters and completed her high-school education. During this four-year stay in New York, Eleanor maintained correspondence with Charles Parker and a commitment to marry that the two had made in 1937.

In the summer of 1941 Eleanor returned to West Virginia to wed "Rex" Parker (Charles's newly adopted stage name), who was then playing mandolin with the Holden Brothers (Fairley and Jack). On August 31, 1941, there was to be a WJLS sponsored country-and-western music show at the Raleigh County Ball Park in Beckley, featuring Dixie Lee Williamson (Molly O'Day), Lynn Davis and the Forty-Niners, and the Holden Brothers. It was decided that Rex and Eleanor's wedding would take place as part of that afternoon's festivities. "We exchanged marriage vows over WJLS," Eleanor told her Beckley audience on August 25, 1990, "in the back end of an old cattle truck."[14]

79

After their marriage Rex and Eleanor quickly developed their own performance routines and a collection of country-and-western numbers, moved their center of "radioland" activity to WHIS-Bluefield, and rapidly rose in popularity in the West Virginia counties of Mercer, McDowell, Wyoming, Raleigh, and Summers (with devoted fans even in Mingo, Logan, and Boone counties). In addition, they frequently played to schoolhouse, theater, union hall, and ballpark audiences, not only in this southern area of West Virginia but also in Virginia and Kentucky. For almost twenty years they were the featured artists at WHIS, broadcasting as many as five shows a day, enjoying the long-term sponsorship of Tomchin Furniture Company of Princeton, and performing with the popular Lonesome Pine Fiddlers and their own group, the Merrymakers.

According to country-music scholar Ivan M. Tribe, during these World War II years country-and-western music had not yet become thoroughly centralized in Nashville, Tennessee.[15] Radio stations scattered throughout the South and Midwest served as home bases for country singers and musicians who staged commercially sponsored live programs throughout the week, but particularly on Friday or Saturday nights—"The Wheeling Jamboree," "The Old Farm Hour," "The Renfro Valley Barn Dance," and of course "The Grand Ole Opry."

Usually dressed in stylized western garb or exaggerated "hillbilly" fashions, and identified by show names that played on western, country, or mountain themes (the Skillet Lickers, the Border Riders, the Prairie Buckaroos, the Happy Valley Boys, the Melody Mountain Boys, the Clinch Mountain Clan), country-and-western bands entertained radio audiences with a mixture of comedy, romance, patriotism, and inspiration that made the hardships of these war years less burdensome.

When Rex and Eleanor Parker began their radio careers, "bluegrass" had not yet been identified as a distinct musical form. In fact, these western and/or country motifs were to continue to dominate throughout the World War II period—in performance costumes, character names, stage settings, comic-effect makeup, and yodeling tunes.

By 1941 WHIS in Bluefield had become a moderately large center of such country-and-western music activity. Even though it operated at only one thousand watts, this station—by virtue of location, elevation, and sparse competition—played to an expansive audience on both sides of the Virginia/West Virginia border. "WHIS . . . was a real popular station," wrote Garland Hess who worked for the station in

Eleanor and Rex Parker during their WHIS days, 1946. Courtesy of the Parkers.

the late 1930s, "simply because it was the only local station anywhere around. There was not another station in all of southern West Virginia until the spring of 1939, when WJLS in Beckley and WBTH in Williamson went on the air." Hess added that "in the early morning WHIS had regular listeners in Ohio, Kentucky, and North Carolina."[16]

In addition, WHIS had built a strong listening public for live musical programming several years before the Parkers became the station's leading entertainers. This audience had been established largely under the late-1930s sponsorship activities of Bi-Tone Products, a line of nonprescription medicines manufactured by the Jeffries Drug Company of Bluefield, including a "Wonder Tonic," a cough syrup, a cold capsule, a liver pill, and an antiseptic. "By 1937," writes Ivan M. Tribe, "the Bi-Tone Products Morning Jamboree provided two hours of [live] entertainment each weekday morning. Lynn Davis and his Forty-Niners headed the cast."[17]

Bi-Tone Products later went out of business when a new pure food and drug law forced the company to publish, on its products' labels, all ingredients. "When the word 'strychnine' was duly listed," writes Garret Mathews, "the best salesman in the world couldn't sell the stuff."[18]

Sitting on the eastern edge of the Pocahontas coalfields, Bluefield, West Virginia, and her sister city, Bluefield, Virginia, had become a busy supply center for mining operations in Tazewell, Buchanan, Russell, and Dickenson counties in Virginia, and Mercer, McDowell, Wyoming, and Raleigh counties in West Virginia, an expansive area that, during Rex and Eleanor's early years at WHIS, was experiencing a wartime boom. By 1944, an article in *The West Virginia Review* could tout Bluefield, "Capital of the Pocahontas Coal Field," as a progressive city of 25,600 (including the Virginia side), serving a trading area that annually produced 30 million tons of coal and generated an estimated yearly income of $200 million, approximately $69 million of which came from the payroll of the Norfolk and Western Railroad Company, shippers of the vast tonnages of coal that moved through the Elkhorn Creek and Tug River valleys northwest of Mercer County.[19]

Rex Parker avoided World War II military service because of a heart murmur, and thus he and Eleanor were able to benefit both from a war-stimulated economy and the popularity-building powers of the golden age of radio. Frequently on the road, traveling as far as three hundred miles from Bluefield, they staged entertainments before assemblages of farmers, logging and coal-camp audiences, and a great variety of small-town publics, always producing what Eleanor says was a wholesome mix of "silliness, entertainment for the whole family"—picking and singing intermingled with such standard crowd-pleasing events as the ugly-man contest and the fiddling competition.

In the latter happening Rex would challenge the best local fiddler to a battle of bows, with the audience determining the best performance, and the loser condemned to roll a peanut across the stage with his or her nose. More often than not, says Eleanor, the crowd awarded victory to the local contestant just to see Rex do his peanut-pushing act. Always the good sport, Rex would make the most of these moments, entertaining audiences perhaps as much with this routine as with his mandolin or banjo performances. "Folks wouldn't laugh at such stuff today," she adds, somewhat wistfully. "We played dances, too," she now admits with some reluctance, "but we always liked the stage shows."

The lumber and coal camp crowds could be rowdy, but usually there were enough women and children present to temper the moods of these occasions. In addition, Rex and Eleanor always played a number of hymns and patriotic songs, both of which could be used to sober a World War II audience.

Travel was made difficult during those war years simply by the fact of gasoline and tire rationings, and by the absence of any new cars for civilian ownership; but Rex developed the resourcefulness and skills of a shade-tree mechanic and kept a 1930s-model Willys station wagon functioning. Eleanor recalls trips made by their entire band, instruments and all, in this vehicle, pooling gasoline quotas and switching tires from one car to another.

Rex and Eleanor Parker speak of these years with a great deal of warmth and pleasure, but also with a sense of loss. These were the youthful years, the high-spirited years, the innocent years, the patriotic years, the years during which rural and urban audiences alike could be entertained by routines filled with clownish frivolity. Country-and-western entertainers could mount a stage and use their musical numbers and nonmusical acts to swing sharply from religious moods, to romantic moods, to comic moods, to patriotic moods, and always receive an enthusiastic audience response.

"It seems like we all liked simpler things then," says Eleanor, "and that everybody loved Rex and Eleanor."

Today when she uses their names in this way the image thus created is of a tightly meshed unit, a oneness that was and is. For fifty years it has been "Rex and Eleanor," with no separation in the concept, and with no emphasis on a last name to suggest formality.

These were also the years during which the Parker family began to expand. In 1942 Eleanor gave birth to a daughter, Conizene, and in later years the family grew to include a son, Charles Clarence, and a second daughter, Rexana Delleen. As these children developed they were included in the family performing troupe, with Conizene becoming known at the early age of three as the "Singing Sweetheart of the Airways."

There was a fourth child, a girl, Ola Marcina, born July 13, 1945; but this infant lived only six months, her sudden death leaving Eleanor inconsolable for many months. Eventually Eleanor was able to pour her sorrow into a poem about the little girl, still later setting the words to music. "For the longest I couldn't sing it," she says, "but then God allowed my song to ease my pain."[20]

Our Baby's Memory
by Eleanor Parker

February the fifth in 1946, at the close of the nine o'clock hour
Sorrow had come that day to our home
And the memory will linger forever
Our baby girl had been well that day, and nothing was
    seeming to hurt her
But God saw to it and took her away,
And her memory will linger forever.
She stayed with us only six short months, so short they
    seemed only a day
But now that she's gone a crown she has won
And I hope to meet her someday
She rested so peaceful in her casket snow white,
A smile on her face as when living
God gave her to us as a rosebud on earth, to bloom as a
    flower in Heaven.
Oh Lord, how I long to hold her once more, and feel her
    small hand on my face
Tho I know this can never be no never
Yet her memory will linger forever
Her dress snowy white with ribbons so pink, her cheeks
    like the first rose of summer,
Her big brown eyes closed in peaceful rest,
Her memory will linger forever.

Such tiny pink feet and small cuddly hands, a picture
   I'll never forget
She was taken to join God's heavenly band
It seems I can still see her yet
Dear God up in Heaven please hear my prayer, guide me
   thru life day by day
Help me to know I'll meet baby up there
As I trod the golden narrow way.
Dear baby girl oh how we miss you, forget you never can never
Yet we'll hope with a prayer to meet you up there
And your memory will linger forever
Her dress snowy white with ribbons so pink, her cheeks like
   the first rose of summer,
Her big brown eyes closed in peaceful rest,
Her memory will linger forever.[21]

"Most groups of the area," asserts Eleanor, "kept moving on. But we just stayed put in West Virginia, playing to the folks who loved us."

Indeed, a wide array of musicians touched base at one time or another in the Bluefield, Princeton, and Beckley area before "moving on" to country-and-western music centers that provided greater opportunity for a national following—Tulsa, Dallas, and Nashville, among others. The experience of the husband-and-wife team of Lynn Davis and Dixie Lee Williamson (soon to become nationally known as Molly O'Day) became a case in point, leaving WJLS, as they did in 1942, first for WAPI in Birmingham and then for a fifty-thousand-watt station, WHAS in Louisville.[22]

In the early fifties the Parkers did take a brief trip to Nashville to record four songs for the Coral label,[23] but for the most part there was no strong move on the part of the Parkers to climb to some loftier level of stardom. Instead, the life of the Parker family became a routine of commercially sponsored radio broadcasts to the southern region of West Virginia and the westernmost counties of Virginia, augmented by constant travel to meet show commitments.

"When we started we charged only fifteen cents a ticket," says Eleanor, speaking of their theater, union hall, and schoolhouse performances, "but we later had them pay as high as fifty cents."

These receipts, plus their radio commercial revenues, produced a reasonably comfortable living for the family, so they continued to "stay put," and the "they" who "love Rex and Eleanor" have remained in these performers' minds as an unchanging and dedicated body of supporters, still out there, still writing them letters, still telling them

Eleanor, Rex, and Conizene, WOAY-TV, Oak Hill, West Virginia. Courtesy of the Parkers.

how they remember the weekday schoolhouse shows, the larger weekend theater performances, or the still larger bluegrass festivals. It is important to this aging couple that there are Rex-and-Eleanor fans who both remember the glory days and still appreciate these performers' talents. Furthermore, Eleanor has saved enough letters to lend some credence to this vision of a group of followers who were dedicated to the Parkers and their children.

In the early 1950s, when Rex and Eleanor performed two evenings a week on WOAY-TV in Oak Hill, West Virginia, they received numerous letters from viewers who had followed their career since that start in 1941 at WHIS in Bluefield. One letter contained a lengthy poem written by Arnett and Ellen Hamrick of Valley Head, West Virginia, verses that capture not only an indication of the level of fan commitment but some understanding of the sociocultural status of these fans. The references to "Granny" relate to a grandmother-and-her-rocking-chair act that Eleanor regularly performed, playing for both humor and inspiration. The Parkers' TV show was also titled "Songs of Salvation." Aids Discount Center was a sponsor.

At seven thirty on Friday night, we turn the T.V. on,
To get Rex, and Eleanor Parker, who puts in our heart a song,
We love the way they sing and play, and talk about the Lord,
They truly have a gift from God, as it's written in His Word.
We love to hear Sister Conizene sing, she too, is inspired by the Lord,
She starts a song, then all join in, and sing in one accord,
We know they've blessed many weary hearts, for their
    program carries far,
And me,? I get close to heaven, when Rex plays the old guitar.
Many songs he sings, I dearly love, like, Silver haired Daddy of mine,
And when Granny gets in her rocking chair, that's when
    he really shines,
It brings back fond memories, of our life, many years ago,
When we used to see our own Grandmother, rocking to and fro.
I believe that all Grandmothers, who rocks in her rocking chair,
For all the trouble she went through, she'll have a Mansion up there,
And when the shadows of night come down, on the last Earthly day,
We'll thank God for songs of Salvation, that Rex and Eleanor
    used to play.
Heaven will be ready for us all, even before, up there we enter,
Many will credit their Souls being saved, to Aids Discount Center,
For they kept Rex and Eleanor on the air, no matter what the cost,
Without them, on Judgment day, many Souls could have been lost.

We're grateful for the songs of Salvation, which I know
   the Lord has given,
So that some who are lost in Sin, might gain a Home in Heaven,
And the sponsor who pays for time on the air, when their race is run,
Will be with Rex, Eleanor, and Conizene, when Jesus says,
   "Well Done."

Arnett and Ellen Hamrick
Valley Head, W.Va.[24]

Through the years, Tomchin Furniture Company remained their most faithful radio sponsor, but Rex and Eleanor promoted a wide variety of products in the 1940s and early 1950s, including Hadacol, a Louisiana-produced tonic that while promoted as a miracle cure for many ailments apparently possessed little virtue other than its high alcohol content. "They mixed it up and didn't know what is was," now jokingly declares Eleanor, "but they *had a call* it something."[25]

Today Rex and Eleanor are not proud of some of the product promotions they broadcast in those early years, but advertisements of this nature were abandoned once they adopted their current religious commitments. As previously noted, that change in their directions began in 1959 at the close of a period that apparently represented a psychological downturn in their lives.

Neither Rex nor Eleanor is comfortable discussing this period in any great detail, so I am forced to surmise about some forces that may have been at play in their lives. First, by the middle of the 1950s the halcyon days of radio were over. Programming had changed, bringing to the airwaves a greater dependence upon recorded music and a reduced dependence upon live groups. Television was on the scene, and although Rex and Eleanor made their adjustments to the new medium, staging shows on WHIS-TV and WOAY-TV, tastes in entertainment were changing. No longer would the ugly-man contests, the fiddling face-offs, and the peanut pushings satisfy audiences grown sophisticated on the more uptown and dynamically varied fare to which they were exposed on network television. Young viewers, in particular, now found Rex and Eleanor old-fashioned, overly sentimental, and even laughable.

The "folk revival," then in its beginning, had not yet galvanized American interest as it would in the mid-1960s, and appreciation of country music was at a low ebb, due in large part to the rising popularity of rock and roll. All of this suggested to the Parkers that the "folks who loved Rex and Eleanor" might be sharply diminishing.

In addition, the strains of road-show travels began to work negatively on the family. Eleanor reports that she would return from performances feeling depressed, knowing "something was missing." The highs she had previously experienced from all exposures to audiences were no longer occurring. On the other hand, the Parkers may have begun to regret their loyalty to the Bluefield/Princeton/Beckley area: Other performers with whom they had played had moved on to national recognition while they had "stayed put."

Then Rex began to drink: How much, I do not know, but apparently more than was comfortable for the family. Finally, Eleanor experienced some lingering depression in response to the death of Ola Marcina.

The mid- to late-fifties were not unrelentingly dark years for the Parkers: A number of things were happening that brought them considerable satisfaction. Their daughter Conizene was then a teenager and performing regularly with her parents, as were Charles Clarence (known to Bluefield-Beckley audiences as "C. C.") and Rexana, the youngest, described in 1956 by a reporter for the *Bluefield Daily Telegraph* as possessing "the coal black hair, piercing eyes, and olive complexion of her mother" and as "a real heartbreaker," "somewhat of a scene stealer in stage shows as well as on television and radio."[26]

Widely known and popular in the Bluefield-Beckley-Oak Hill corridor, the "Parker Family" had become regular guests on the annual Community Christmas Tree Parties sponsored and broadcast first by WHIS-radio and then by WHIS-TV, originally from the Colonial Theater in Princeton, and then from several school auditoriums.[27]

All of this recognition for the family, however, did not bring total happiness to Eleanor. "For years," she said, "we traveled from place to place entertaining the world, trying to find satisfaction in our lives. We enjoyed meeting people and entertaining them, but this pleasure was only for a few hours. When we would get back home, we would become depressed and we didn't know what was ailing us."[28]

Eleanor was the first to find her answer for these problems in religion. Brother Ernest Barley, a Freewill Baptist radio preacher with whom Rex and Eleanor had become acquainted in their numerous broadcasting involvements, played the role of spiritual counselor to Eleanor, and she reports that one evening she "prayed through," pouring out her emotions to her God and this impassioned exhorter for an "old time, sin-damning, 'Come to Jesus,'" Baptist gospel. "The experience was so good for me," she says, "that it eventually helped Rex and the children."

Rex's drinking difficulties apparently ended, and the family coalesced around new sets of motivations. The hymn singing that once had been a peripheral feature of their performances now became the central focus of their radio and TV shows. Soon Eleanor began to use a statement she still employs when playing for church audiences: "Back when we were playing the schoolhouses and theaters they used to call us hillbillies. Now we're just holybillies."[29]

In the summer of 1959, these performances for church gatherings apparently began to be the central focus of the Parker family. A 1959 newspaper article announcing a Parker Family appearance at the Methodist church in Rhodell, West Virginia, included the following: "Their home is at Lerona in Mercer County, but they travel throughout several surrounding counties, playing and singing for church programs, revival meetings, and for tent meetings during the summer months."[30] No longer were the Parkers just entertainers: They were entertainers with a gospel mission.

Eleanor responds to her moments of grief and her moments of joy the same way: she writes songs about them. So just as she had done upon the death of Ola Marcina, Eleanor celebrated her new life in religion by composing a hymn she and Rex later included in their small publication, *Your Favorite Hymns*.

I've Answered God's Call
Testimony in Song—Written by Eleanor Parker

For years I had been a lost soul in sin
Pretending to be happy each day
But the greatest of all, I've answered God's call
And I'm changed in every way
Satan can no longer tempt me now
My heart's with God I know
Tho once I was lost now I carry the cross
And His blessings upon me bestow.
When I am troubled and dark in the way
I steal away in prayer
I know God is watching and guiding me on
He knows my every care
It seemed so hard to accept thee my Lord
I knew not which way to turn
But now that I've changed from my sins I'm cleansed
And His love in my heart will burn
I'll sing His praises to all that hear
And trust that you'll listen to me

Take Christ in your heart from your sins depart
And your life a blessing will be
I pray that thru me some sinner will heed
And answer God when He calls
Don't go on in sin lose your soul in the end
For God's love is greatest of all.[31]

Rex Parker also composed a song to celebrate his own liberation, in this case from alcohol:

There's Better Things In Life Than Drinking
Written by Rex Parker

You may think you're having fun while drinking
You're only wasting your precious life away
The Devil takes a hold and that's to doom your soul
But there's better things in life than drinking.
CHORUS:
There's better things in life than drinking
Oh come to Jesus today
He'll save you from your sins and reward you in the end
There's better things in life than drinking.
You could go to church each day and learn to humbly pray
And God would fill your heart with joy
He will shower you with love, from Heaven up above
Cause there's better things in life than drinking.[32]

This conversion to a fundamentalist religious life paralleled, to a large degree, a slightly earlier set of circumstances that influenced the careers of Lynn Davis and Molly O'Day: In 1950 Lynn and Molly converted to a Church of God (Cleveland, Tennessee) faith and began to commit their lives to the ministry of this denomination.[33] Since Rex and Eleanor viewed the Davises as mentors, this earlier conversion may have exerted an influence on the Parkers.

Rex and Eleanor's separate religious experiences, however, did not close out their involvements with commercial broadcasting. They continued both their radio and TV work, staging weekly shows on WOAY-TV, Oak Hill, through the 1970s, but with their road-show performances being replaced by appearances at church sings, homecomings, memorials, and the like. In addition, for several years they linked up with Bill Monroe's bluegrass festivals at Bean Blossom, Indiana, and Jackson, Kentucky, featured as gospel artists and apparently also serving the Sunday morning spiritual needs of these

Eleanor, Rex, and Conizene in a 1960 photograph Eleanor labeled "After Our Conversion." Courtesy of the Parkers.

events. Eleanor reports that at one such Sunday-morning bluegrass festival thirty-seven individuals came forward to profess their faith.[34]

By the 1980s, however, Rex and Eleanor Parker found their performance career limited to religious radio broadcasting and to their frequent involvements with special church activities. For several years during this period they performed only on WAEY-AM, Princeton, but recently they have reestablished a Saturday broadcast on WJLS-AM, Beckley.

The glory days of "radioland" stardom, however, are gone. As noted earlier, during those World War II years at WHIS-Bluefield the Parkers had been public personalities of some importance to southern West Virginia, southwest Virginia, and parts of eastern Kentucky, possessing a following that made WHIS and Tomchin Furniture Company proud of their associations with the couple. Held in such high regard, the Parkers received the best WHIS had to offer in terms of attention, airtimes, and production facilities.

That is not the case today, either at WAEY-Princeton or WJLS-Beckley, where FM sister stations have become far more lucrative and widely listened to than these AM systems over which Rex and Eleanor broadcast. For those broadcast periods when simulcasting is not the mode of operation, there appears to be a decided lack of management attention to the AM programming. For example, station managers seem unconcerned about the highly imperfect studio environments within which Rex and Eleanor operate. During the four "Songs of Salvation" broadcasts I visited at WAEY, the following studio problems were evident, at various times: (1) so many fluorescent lights were completely out or just barely flickering that Eleanor could not read her notes on her sponsors, (2) communication with the main studio did not exist (forcing Eleanor to step into the hall to hear the prerecorded introduction and segments of prerecorded music), (3) the soundproofing of the studio was severely compromised by an opening created for a door that had not yet been installed, and (4) the high-school-aged disc jockey at the AM controls kept forgetting to throw a switch before cueing a record or tape, thus allowing the sound to come into the studio where Rex and Eleanor were singing. At WJLS-Beckley, the Parkers performed in a studio that had only one microphone, and it would fall from its stand if you did not touch it exactly right.[35]

Rex and Eleanor Parker, WJLS, Beckley, West Virginia.

Church Sings
The Parkers at the Lord's Full Gospel Mission

Before describing the "Songs of Salvation" broadcast over WAEY, I want to introduce readers to that part of the Parkers' lives that relates to their performances at small Appalachian churches. On August 19, 1990, the Parkers performed at a church "sing" conducted at the Lord's Full Gospel Mission just off Route 621 in Pilot, Virginia, a rural community near Christiansburg. Earlier the Parkers had broadcast their regular 8:00–9:00 A.M. show over WAEY. At 11:30 I met them at Kinney's Restaurant, one of the businesses Rex and Eleanor regularly promote during their Princeton broadcast. I then trailed the Parkers' vintage Cadillac in a briskly paced trip across three counties, following Highway 460 from Princeton to Blacksburg to Christiansburg, and from there on to Pilot, Virginia.

As has already been established, Rex and Eleanor are still very much "on the road," traveling not to those lively and sometimes raucous stage performances at schools, theaters, union halls, or ballparks, but to small churches tucked among the Virginia and West Virginia hills. There they entertain small audiences of unsophisticated fans, gathered for revivals, homecomings, family reunions, memorial services, decoration days, church raisings, or special sings.

Such audiences usually include a number of aging individuals who remember Rex and Eleanor during the WHIS, WJLS, or WOAY-TV "glory days." Judging by what I heard at Pilot, Virginia, these older followers place the Parkers in a memory frame that dates back thirty, forty, or fifty years, perhaps in the process seeing these performers as young as they (the aging audience members) occasionally still see themselves.

On this particular Sunday, the Parkers' travels took them to the rural church of Brother James R. Kittinger, an independent Pentecostal preacher who has gathered a small group of followers around activities of the Lord's Full Gospel Mission. This concrete-block house of worship, crowned by a short steeple that breaks the boxiness of the structure, sits at the end of a gravel road that exits from Route 621 in Montgomery County, Virginia.

A recently painted sign welcomes the visitor, announces service times, declares the church's name and pastor, and proclaims the day of founding, May 22, 1987. The building rests in a slight depression between a wooded knoll and a rolling expanse of pasture—a tranquil

The Lord's Full Gospel Mission, Pilot, Virginia.

setting that represents well the picturesqueness of Virginia's western Blue Ridge foothills.

The church facility—a new structure—is small, consisting of the sanctuary, two Sunday-school rooms, and two restrooms. Its meeting space accommodates only seventy-five or eighty worshipers; nevertheless, the membership of this "Full Gospel Mission" can fit in that space with seating to spare.

Rex, Eleanor, and I arrived at the Lord's Full Gospel Mission at approximately 1:00 P.M., just as the gathering of only sixteen worshipers was about to begin an after-service dinner on the grounds, served around a long picnic table set inside a screened-in shelter some one hundred yards from the church. The table was spread with a wide variety of meats, vegetables, and desserts, home-cooked, full-flavored, heavy with taste-giving fats, and far more than was needed to feed this small collection of Pentecostal churchgoers.

I was immediately welcomed by the small congregation of celebrants, having come with Rex and Eleanor and therewith assumed to be properly supportive of all values vital to this gathering. Brother Kittinger, a stocky, friendly, passionate man in his mid-forties, was particularly quick to greet me, shaking my hand and embracing me as if I were a regular member of his flock.

An uncomplicated man with minimal education, Kittinger communicates his commitment to his calling via energy, enthusiasm, and frequent statements of acquiescence to "God's will," making it clear, however, that this "will" includes the working out of a plan for this modest congregation of Pentecostal Christians.

Shortly after our initial meeting, Brother Kittinger learned from Eleanor who I was and what I was about. His response was to hand me a business card—"Rev. James R. Kittinger. Will schedule revivals and singings. Lost souls are what we are searching for."—and to tell me that after our meal he would show me around the church grounds and tell me about his plans for the "Mission."

True to his word, we later took that walk, with Kittinger revealing how special this plot of land was to him and how deeply rooted was his dream for a small but strongly motivated body of churchgoers who would anchor their spiritual lives to his ministry. He even took me back into a wooded area where he had built a small shed, lifted some eight to ten feet off the ground by sturdy poles. He sleeps there occasionally, he said, after he has worked long hours on the church, rather than driving back to his home.

Brother James K. Kittinger lives in Newport, Virginia, a diminu-

97

tive community northwest of Blacksburg on Highway 460, but he has established his "mission" some thirty miles southeast on land that has been in the family for many years. His long-range plans include the construction of his parsonage on a cleared knoll some three hundred yards from his church, transforming this small spot of Virginian soil into an enclave of piety. Kittinger's dream is comparable to that of Brother Johnny Ward of chapter 2, differing only in that it involves a church and a parsonage rather than a revival tent and a bus. While Kittinger sees himself as a pastor, Ward sees himself as a traveling evangel.

The interior of Kittinger's church gives evidence of loving care and money-saving inventiveness, equipped as it is with a mixture of homemade and salvaged furnishings. The pulpit is one of the homemade items, suggesting an unplanned coming together of disparate elements to forge a thing of ultimate pride in craftsmanship. Constructed of lightly lacquered hardwood panels embellished by a variety of moldings, this structure represents Brother Kittinger's seat of spiritual influence, the place from which he meets his call to exhort.

An unadorned wooden cross is set in relief against the congregation side of the pulpit's cabinet, while the gilded message "Jesus Saves" graces the face of the lectern segment of this structure. Underneath this ornamental declaration, sitting on the mantel of the lower section, a Dove-of-Peace figurine nests in symbolic serenity. Finally, to the right and left of this dove, affixed to the outer wings of that mantel, there are two glass balls—the type that when turned over simulates a snow scene—each containing a romanticized three-dimensional representation of a Holy Land pastoral.

Two microphones stand beside the pulpit to the congregation's left, and one swivels from the left wing of the pulpit's lower structure. These mikes feed to an amplifier standing against the back wall, a unit which in turn is wired to two large speakers secured in the upper corners of this wall. All of this guarantees a volume of celebratory sound far larger than is needed in this small worship space. This is a Pentecostal church, accustomed to an exuberance of expression, as we have already seen demonstrated in Sister Kathy Benfield's church, and as we will see again in chapters 4 and 5.

Immediately behind this pulpit hangs a large, inexpensive, painted-on-black-velvet rendition of Jesus as shepherd, leading his lambs to water. This work is flanked, on the congregation's left, by a wooden plaque depicting steepled hands, above which is printed "How Great Thou Art." To the painting's right hangs a smaller pic-

ture of Jesus (this time as the crucified Christ), again on black velvet, set in a gilded rococo frame, and illuminated by a hooded light.
During evening services Kittinger occasionally turns out all other
lights, thus leaving this crucifixion image sharply focused.

The cumulative effect of all these inexpensive religious artifacts is
an image of cherished tawdriness. Like every other feature of this
small church, these icons of theological simplicity assume a collective
sense of loving dedication, the abstracted spirit of Brother James
Kittinger's mission of gathering together, souls and sacramentals, all
parts of the dream.

In front of the pulpit is a fifteen-to-twenty-foot expanse of altar
railing, constructed of two-by-four stock, and upholstered on top with
a deep maroon velvet. Two items sit on this structure: a "Prayer Request Box" and a small model of a traditional, wood-framed, white
church. Congregation seating is provided by several rows of refurbished theater seats, obtained no doubt from a demolished movie
house or school auditorium. Augmented by the presence of the previously mentioned microphones and speakers, these theater seats suggest that performance-space image that is so much a part of many
Pentecostal meeting houses.

A small electric organ and an equally small piano sit side by side
against the upper left wall, both giving the appearance of having been
only recently purchased. These units are balanced on the opposite
side of the pulpit by a section of theater seats that face left toward the
preaching area.

All is neat and orderly, clean and freshly painted, and suggestive of
an abundance of attentive care. This structure literally is "Brother
Kittinger's church," and he puts every dollar he can into the facility's
improvement and upkeep.

The "sing" began that afternoon at 2:30. By then the group had
grown to approximately forty people, some of whom brought musical
instruments. Brother Kittinger opened the event with a verse or two
of Scripture and then a brief sermonette. Then he began to call on various groups who apparently were members of the church. A husband-
and-wife team sang a number of songs, accompanied by the guitar
playing of Brother Kittinger himself. Then Kittinger was joined by
his wife and daughter, again for several hymns. My favorite was a constantly smiling, almost toothless man, with a harmonica harness
around his neck and an acoustic guitar in his hands, who rendered,
among other tunes, a loud, twangy, and joyous version of "The Lily of
the Valley," pouring an immense amount of happy and unvarnished

The singing service at the Lord's Full Gospel Mission. Brother James Kittinger, *at left.*

passion into his performance. Neither his singing nor his playing (harmonica or guitar) possessed any degree of polish, but his exuberance was impressive.

After forty-five minutes of musical expression from these groups within the church, Kittinger introduced Rex and Eleanor, who immediately conveyed a sense of on-stage showmanship. Years of performing to schoolhouse, union hall, and theater crowds have established a style that is still there, particularly in Eleanor, and even before a small gathering such as this.

Eleanor immediately began her fast-paced, colloquial banter, moving from chatty humor to religious inspiration with a grace that communicated not only her comfort but her absolute delight in being before this body of old fans, limited in number though they were. At times I caught glimpses of that 1940s performer who learned to use her wit, waggery, and word play to control rough-and-tumble coal-camp audiences, moving them whenever she wished from spirits of frivolity to moods of piety or patriotism.

The real charm in Eleanor Parker's style is that she does all this with a naturalness that communicates honesty, blending folksy repartee with rhetoric of sincere reverence. Ultimately she sees these performances as having evangelistic missions; therefore, no matter how far she slips into quick and witty banter, her final goal is religious exhortation.

Rex appears responsive to all this, in a very low-key sort of way; but he seldom speaks, leaving all interlocutor functions to his wife. As mentioned earlier, he will smile occasionally, especially at the close of his fast-paced instrumental solos, and he exhibits some expressiveness when he sings. Nevertheless, it is always Eleanor who verbally sparks the show, partnering her faculty of speech with his faculty of dexterous performance.

Rex and Eleanor entertained and inspired this small audience of believers until late in the afternoon. At 4:30 P.M. I was forced to leave the Lord's Full Gospel Mission in order to make it back to Boone, North Carolina, in time for a Sunday evening responsibility, but I departed this scene with some regret: I had found myself falling under the charm of these two performers from the golden age of radio.

Rex and Eleanor performing at the Lord's Full Gospel Mission.

"Songs of Salvation"
WAEY-AM

During the following discussion of the Parkers' WAEY broadcast, emphasis is placed primarily on the commercials, since—as has already been noted—these advertisements make "Songs of Salvation" distinct from all other airwaves-of-Zion programs examined in this volume. At the time of my initial visit to WAEY, the Parkers had four sponsors: Gale's Resale, Kinney's Restaurant, Mills' Market, and Magic Mobile Homes, each apparently supporting a quarter of the broadcast's one-hour airtime. On that particular morning, however, Eleanor devoted over twenty minutes to Mills' Market and slightly less than ten minutes to the show's newest sponsor, Magic Mobile Homes. I saw no evidence, however, that Eleanor was even taking note of these time allocations. The show's structure seems to be relatively free of such restraints.

Every Sunday morning at eight o'clock listeners to WAEY-AM will hear the following program introduction, played as a voice-over against a fast-paced mandolin tune:

It is with great pleasure we now offer you "Songs of Salvation," with Rex and Eleanor Parker. The Parkers pray that our time we spend together will be a blessing and inspiration to all, and if just one soul is delivered from darkness our time is well spent. If you would like to help Rex and Eleanor in their quest to find lost ones, write Rex and Eleanor Parker, P.O. Box 39, Lerona, West Virginia 25971. We will give the address again at the end of the program. And now, Rex and Eleanor Parker.[36]

Immediately on the heels of this announcement either Rex or Eleanor opens the program with a prayer, immediately followed by the program's theme, the lyrics of which go like this:

He can,
and he will,
save your soul
from the burning fire,
if you'll only
put your trust in Him.

Now Eleanor reads a brief verse of Scripture, speaks a personal welcome to the radio audience, and segues into the first commercial of the program, housing this introduction in religious phrasings that suggest—and even claim—spiritual blessings for listening and for buying.

Each commercial segment is preceded by a musical jingle. Eleanor writes these doggerels and sprinkles them with her West Virginian rustic wit. More than any other feature of the broadcast, these jingles provide "Songs of Salvation" with its 1940s flavor.

Delivered from a collection of notes she places on the studio lectern, some of which invariably become mixed or fall to the floor, Eleanor's commercials are always informal, conversational, and heavily anecdotal in tone, appearing to have no predetermined format or time frame, except when she falls into a spiel that obviously has been set to memory.

At times, as was the case during the April 16, 1989, broadcast of "Songs of Salvation," one of Eleanor's commercials will appear to end, only to experience a rebirth later in the hour. On that April broadcast, for example, bits of the Mills' Market commercial were scattered over about twenty-five minutes of the show, breaking out between songs and between at least three segments of discourse concerning individual supporters of the program who should be remembered in prayer.

Although spiritedly delivered, these advertisements are simply talks to the radio audience—down-home, front porch, and neighborly, occasionally exhibiting touches of tongue-in-cheek frivolity, and always demonstrating Eleanor's fascination with the delicately turned lyrical phrase. For Eleanor words have a poetry and finesse that must be highlighted just so by a lilting delivery that accentuates the natural rhythms of Appalachian speech. Consider as an example of all the characteristics mentioned above the rhetoric of Eleanor's opening to the Gayle's Resales commercial:

> Greetings once again in the precious name of our Lord and Savior Jesus Christ. Another beautiful Lord's day God has spared us, this side of eternity, to visit with our wonderful friends along the airways, and we're certainly trusting and praying that your hearts will be blessed as we bring you the message of salvation, words from our sponsors, and acknowledgments of our mail and requests for this week.
>
> Our very first selection we'd like to send out to the folks who visited Gayle's Resales this past week, and we thank you so much for mentioning the broadcast to Bob Willby. I stopped to check with him yesterday, and he said he was real pleased with the results of his commercial. So thank God for that. That's Gayle's Resales, 604 Mercer Street, here in Princeton, and also at 103 Waters Street at Peterstown, West Virginia.

Here Eleanor drops into a delivery mode that is somewhat faster and communicates the idea of an established spiel. In doing so she plays more strongly on the lilting cadence her words always possess, particularly on such phrases as "knickknacks or notions."

Now they have appliances, clothing, baby needs, and furniture, an indoor yard sale throughout the week, Mondays through Saturdays, from 10:00 A.M. to 6:00 P.M., and when you're in need of furniture, clothing, knickknacks or notions, bargains at extra low prices, come and browse around. Because you're bound to find just exactly what you want and what you need at the right price at Gayle's Resales, 604 Mercer Street, Princeton, and also 103 Waters Street, Peterstown, West Virginia.

Now for this week his special is a five-drawer wooden chest which is just right for a child's room to put away their clothing and whatever you might want to store in those chest of drawers. They are originally priced at $69.95, and mentioning the broadcast for this week alone you can get this particular chest for just $59.95. Tell them Rex and Eleanor asked you in, getting yourself a 10 percent discount on the price of any of the items you might choose. Gayle's Resales. Tell Bob that Rex and Eleanor asked you in. We'd sure appreciate it, and I know that God will bless you for it. That's for sure.

"God will bless you for it. That's for sure." These words demonstrate that in Eleanor's mind there is no troublesome conflict between spiritual messages and commercial messages, as long as she can believe in the inherent goodness or worthiness of the advertiser and product or service. "I'm not going to advertise something or someone I can't believe in," she says, and presumably the frequent visits she and Rex make to Gayle's Resales satisfy her that this is a business concern with an acceptable record of good conduct. It would make a difference, Eleanor told me, were a number of listeners to write the couple and charge that a particular commercial institution "did them [the listeners] wrong."[37]

Messages of selling, therefore, mix appropriately with messages of salvation if mankind is well served in the process—commerce and Christianity, not one and the same, but working hand in hand. By Eleanor's judgment, the placement of a man within an environment of beer, wine, and lottery tickets, among other possible social evils, does not result in his being well served. If, on the other hand, a father is able to acquire an inexpensive chest of drawers for his child's room, a family might end up being very well served. Furthermore, it is not without purpose that Eleanor frequently mixes in her free-form commercials images of "family," "community," "friendship," "thrift," "golden-rule responsibility," "service to mankind," "loyalty," "human commitment," and a host of other values she promotes through her less secular activities and rhetoric.

One of Rex and Eleanor's long-time sponsors is Kinney's, a restaurant just east of downtown Princeton on Highway 460, specializing in chicken or steak lunches/dinners and in their generous breakfasts. Kinney's perhaps does not deserve the "greasy spoon" metaphor, but it is still not the place one would want to patronize when on a low-fat diet.

When Eleanor launches into an advertisement for Kinney's, she spends at least five minutes describing meals she has chosen to emphasize, throwing in a plug or two for a couple of items she and Rex market at Kinney's. To hungry listeners these lengthy depictions of food can be quite persuasive; and, according to David Price, the restaurant's manager, Kinney's is more than satisfied with the influence the Parkers have on business.[38]

Following the musical jingle Eleanor has composed for their Kinney's promotion, the WAEY Sunday morning audience is apt to hear something like the following largely improvised message. Readers should imagine a folksy but sprightly delivery that occasionally plays teasingly on certain rich images of savoriness. When the latter occurs, Eleanor is always smiling, knowing full well the impact of her words on appetites.

> If you're traveling along Highway 460, east of Princeton, and you find yourself taking the hungries this morning, then I extend a cordial invitation to you that you might visit out at Kinney's. That's Route 460 East into Princeton. Because right now he'll be serving a great big ol' platter breakfast for you, and it's a country breakfast that's country enough to tease and tasty enough to please. You better believe it!
>
> You get your breakfast platter with a great big ol' brown hoecake, a bowl of sausage gravy—brown gravy. Oh, it is lipsmackin' good! Also your two big slices of bacon or your two big brown sausage cakes.
>
> You get one egg. Did I say the cackleberry, Howard?

She breaks in here to speak to J. Howard Kress, who tells her she has not mentioned the egg; whereupon, Eleanor continues:

> And a big ol' country cackleberry, and your fresh coffee. You get the regular or the Sanka, either one. Free refills on the coffee. Two slices of red-ripened tomato, and they are nice and juicy, and, boy, I tell you they are good, good, good!
>
> Talk about getting hungry! I already am, and I just started talking

about them. But you get your breakfast out at Kinney's this morning. And their breakfast hours, by the way, on Sunday are from 8:00 to 11:00 A.M. Breakfast hours throughout the week, Mondays through Saturdays, from 7:00 to 10:30. The breakfast platter, a dollar and ninety-nine cents!

Be sure to mention Rex and Eleanor when you go in, because this is most encouraging to David Price to let him know that you do listen to the broadcast and that when he helps to pay for a portion of the airtime that money is well spent.

Later in this commercial Eleanor provided additional information concerning how the Parkers make their money. She told the listeners to mention to the manager of Kinney's that their patronage was in response to the broadcast.

For when you do, dear friends, he's gonna take your ticket, put our name on the back of it, and at the end of the month he tallies up the total. And whatever it comes to, he will give us five cents extra on the dollar. And we do appreciate that so very, very much! Our sponsors that pay for a portion of the airtime, then our friends that love us. You send in your love gifts along with it to keep the broadcast airtime expenses up to date.

And we want to mention our giant-print Bible that David has on— He doesn't have them on display, but he has them there. You ask for them if you want to see them. Twenty-two fifty for the giant-print Bible, or six dollars for the number 101 cassette tape of yours truly Rex and Eleanor singing. And J. Howard Kress is playing on there, Brother Ira Chaffins, and also our daughter, Conizene, is helping.

These last remarks were supposed to signal the young man in the studio to play one of the songs from the tape Eleanor just mentioned, "Rex and Eleanor Parker Sing Old Time Pentecostal Revival Songs," six traditional tunes with lyrics rewritten by Brother Dallas Turner, a Pentecostal preacher of Reno, Nevada. The young man on the control board had become distracted and missed the cue; thus Eleanor had to tell him—through the open mike and with the tone of a mother's firmness—that all was not going well. The song played from the control room that morning was "Gospel Wildwood Flower," sung to the tune of the traditional folk song, "Wildwood Flower."[39]

Following this tune, or some other number from "Old Time Pentecostal Revival Songs," Eleanor traditionally delivers her Mills' Market advertisement, which again opens with a musical jingle and develops somewhat as follows:

Yes indeed, the folks at Mills' Market truly aim to please. They're located just outside of Princeton, next to the last stop light on the right as you go towards Bluefield. You just can't miss it, for sure. Mills' Market. We invite you to stop in, because they do have a nice large shopping place, with plenty of free parking space. And you're always going to find a large display of seasonable fruits, fresh garden vegetables, a stock of staple goods, dairy products, a bakery shelf with all kinds of fresh pastries, breads, and snacks that'll tickle your taste buds. And that's not all. Mills' Market also carries a good supply of drugs and cosmetics, a beautiful variety of spices and herbs for seasoning or canning. And in fact there is more in the store than meets the eye. So if you need something, and you don't see it on display, dear friends, just call for it by name. Because more than likely the Mills boys will have it, and if they don't they'll do their level best to get it for you. Just like I say: If you want it, name it; if they've got it, you'll get it; if they ain't, they'll get it; and, of course, if they can't, then you just forget it, because you probably don't need it nohow; and that would be an absolute fact!

Be sure to see Buddy Mills, a long-tall-drink-of-water storekeeper, or any of the Mills clan that might be there to wait on you. There's Corky, Pat, Brenda, Bobby, or Sunny, a fine family of cornbread-fed friends who really appreciate every, every customer. And they have your best interests at heart.

Now convenience to you, the customer, is truly their first concern. They're big enough to serve you, yet small enough to know you, and they've been in the marketing business now for well over forty-nine years, almost a half a century. And they're mighty proud of their record of establishing a wonderful reputation with their many, many friends over these many years.

Now Mills' Market really appreciates each and every customer who patronizes their business, and we want to tell you folks, that when you go in, mention to Buddy Mills or his brothers, or whoever might be there in the establishment, that Rex and Eleanor asked you to do so. Because you'll be doing yourself a great favor as well as us, and at the same time encourage our sponsors to keep us on the air for God's glory.

Be sure to tell Buddy Mills when you go in today that you heard the broadcast. Okay? We sure would appreciate it.

As indicated earlier, the Mills' Market commercial stretched over a considerable length of airtime, broken by—among other things—a number of humorous interchanges between Eleanor and J. Howard Kress. When Kress is in the studio, which is not always the case, Eleanor uses him as an end man, the two playing off each other to provide the broadcast's moments of lightheartedness. Rex apparently has never been able—or willing—to play that role.

When it became time for Rex, Eleanor, and J. Howard to play another song, Eleanor had trouble finding the lyrics: It had become so
dark in the studio. She took this opportunity to scold the station for
the studio conditions, expressing the reprimand in such a way, however, that it would have been difficult for the station's management to
respond defensively. "The light has gone out here in the back as
usual," she said, without a harsh tone to the remark.

> I said it seems like the Devil really tries us every Sunday morning
> 'bout this time. If it's not the recordings, if it's not the speaker in there,
> it's just about got to be the lighting. It was clear out when we walked in.
> It's come on and it's about half lit. So we'll do the best we can. Got to
> almost memorize the words to these songs. It's been a long time since
> we sung them.

All this became a fitting, although accidental, introduction to the
group's next hymn: "If the Light Has Gone Out in Your Soul."

After two more hymns and another round of comments about Mills'
Market, the next segment of the program was devoted to recognizing
a large number of individuals: some who were "sick and shut-ins,"
some who had contributed to the broadcast, some who had been long-
time friends of Rex and Eleanor, some who had gone through surgery
or other medical treatments the past week, and one unidentified "sinner friend" who had written Rex and Eleanor to ask for prayer, contributing five dollars in the process. During this period Eleanor also
acknowledged my presence in the studio, reading my name from the
business card I had presented and communicating to the radio audience her understanding of my airwaves-of-Zion field research.

This moment provided Eleanor opportunity to engage in some
reminiscing, speaking intimately to her "radioland friends" of long
standing, reminding them how long she and Rex—and J. Howard
Kress—had been broadcasting in Princeton, "well over forty-three
years," she said, not counting the earlier and then overlapping years
they had been at WHIS in Bluefield.

> We were here when the old station was up, WLOH. And then, of
> course, it changed hands, and we've been here [WAEY] ever since then.
> And we just thank God for the people we've met down through the
> years. As I told the gentleman here in the studio, a while ago, I guess
> you'd say Rex and Eleanor are more or less a byword.

Then she added, "And we want to be here to Jesus says it's enough,
or until he calls us home. And it's up to you people," she continued, "to

stand by us with your prayers, your love gifts, and to shop by visiting our sponsors, letting them know just what these programs are meaning to you Sunday after Sunday. It sure is a blessing to our hearts."

Eleanor also found opportunity to add that, at WAEY, Mills' Market had been their longest-running sponsor. "Fifteen years they've supported this program," she said, "and they're much appreciated." Later she called out the first names of all the people who currently work at this establishment, asking Jesus to bless them.

After this period of reminiscing and of acknowledging all their friends and supporters, the trio sang "This Ol' House," followed appropriately by Eleanor's commercial for Magic Mobile Homes. As she had done with the earlier commercials, Eleanor again began with a musical jingle. The first two lines of this jingle were sung to the tune of the "Just looking for a home" refrain to the old boll weevil folk song,

> Are you looking for a home?
> No need to roam.
> Say you're paying rent
> With money spent
> And nothing to show for it?
> See Tom, Jack, and Bob
> At their mobile home lot.
> Your drive will be well worth it.
> Special deals today,
> With terms to pay,
> At Magic Mobile Homes.

Eleanor laughed and said she thought she had sung the jingle correctly this time, after not doing so on previous occasions. Later she noted that Magic Mobile Homes was their newest sponsor and that she was still learning all about the "folks" who ran this company. Before she made these remarks, however, she delivered the following:

It is for sure, Dear Friends, if you're paying rent where you are, you know that at the end of the year you've only got a handful of receipts that J. Howard said wouldn't even start a good fire. And that truly is money spent and nothing to show for it.

So if you want to buy yourself a beautiful mobile home we want to say, "You just try all the rest, but you'll settle for the best when you stop in at Magic Mobile Homes, located one mile from Bluefield, West Virginia . . . ." And on the lot today and every day you're going to find a mobile home that is suitable to your liking, at terms that are customed to fit your needs. Easy financing, service after the sale, and free setup and delivery.

This was not all Eleanor had to say about Magic Mobile Homes. The last ten minutes of this April 16, 1989, broadcast were devoted to this new advertiser, with several urgings that listeners just go by and get acquainted with this newest member of the "Songs of Salvation" family. Eleanor wants her radio audience to know her sponsors by their first names, and it is also important to her that very personal ties be established between these sponsors and Rex and Eleanor. "I don't want to speak for someone I don't know and can't feel friendly towards," she says. "That's not the kind of business I want to do." "Tell 'em Rex and Eleanor asked you in," she urges with every commercial, and when she says this she means it in the most neighborly and/or church-fellowship sort of way. If a business joins the "Songs of Salvation" support system, its owners and managers become "Christian brethren" whether they seek that distinction or not: They are helping the Parkers in their "laboring so long."

A Final Observation

The last time I visited the WAEY "Songs of Salvation" broadcast, September 16, 1990, Rex was ill and J. Howard Kress had not come to the station that morning. As a result Eleanor was doing the show all by herself. In addition, that was the morning the studio sound system did not work properly, causing Eleanor to miss cues coming from the main console, operated on this occasion by a young woman recently graduated from high school.

It was not a good morning for Eleanor Parker. To provide variety to the hour-long production, she had asked the young woman on the board to play several of the selections from the "Old Time Pentecostal Revival Songs" cassette she and Rex had recorded for Brother Dallas Turner, the Pentecostal minister in Reno, Nevada. She was trying to fill in with her commercials and a few songs sung just by herself. The only problem was that she could not remain in her studio and know when a recorded song was ending.

For the entire hour Eleanor ran back and forth between the two studios, trying still to keep her composure and her on-the-air enthusiasm. All of this was a far cry from what I imagined her circumstances to have been during Rex and Eleanor's golden days at WHIS, when the two enjoyed the sponsorship of Tomchin Furniture Company, and when life for the Parkers was full of youth, war-years excitement, patriotic songs, show dates and travel, schoolhouse audiences, ugly-man contests, peanut pushings, Hadacol jingles, the old Willys station wagon, the children, and considerable fan admiration. "I will keep going," she told me that Sunday, and I found myself hoping she could.

—4—
Brother Dean Fields, WNKY, and "The Words of Love Broadcast"

I've never been to a better church. It's the love that's here. And I practice tongues. Some preachers don't like that no more. Brother Fields said, "You speak whenever the Spirit calls and whatever the Spirit calls."

Sister Martha Adams

Wright Fork Creek begins somewhere northwest of Jenkins, Kentucky, on the western slopes of Pine Mountain, and flows down through the communities of McRoberts, Fleming, and Neon—all in Letcher County—before it combines with Potter Fork to form the Boone Fork of the North Fork of the Kentucky River. Named after Joel Wright, father of "Bad John" Wright, the legendary lawman and feudist of Pine Mountain,[1] this narrow stream winds through a tight valley that once supplied American industries and homes with thousands of tons of coal a day, the tonnages being snaked out of the hollow on a spur of the Louisville and Nashville Railroad, other lines of which then moved the coal first through Hazard, Kentucky, and from there to Louisville or other shipping points along the Ohio River.[2]

Today the L&N tracks are gone, except for an occasional section buried in the asphalt of a street; and the shaft mines are completely closed, leaving such coal miners as still reside in these communities with the daily task of commuting to active operations elsewhere in the region, primarily to various strip-mining jobs on the Virginia side of Pine Mountain. The three small towns of the Wright Fork region are still there, but only as residual remains of the commerce and industry of livelier days.

As part of the rich Elkhorn field, the initial Neon, Fleming, and McRoberts chain of mines was opened between 1912 and 1914 by Consolidation Coal Company and its affiliate, Elkhorn Fuel Company,[3] with these communities being built as "company towns"—long, narrow, and tightly packed stretches of residential row houses, mining offices, the company store, other commercial establishments, and churches that shared the meager valley floor with Wright Fork, the railroad bed, and a one-lane crushed-stone road. Some homes did creep up the steep slopes of the mountains, two and three levels above the flood plain; but these tended to be the houses of the more-affluent residents—mine managers, merchants, and the like.

Eventually the road was widened and paved, and the original unpainted, company-built houses were replaced by structures somewhat more comfortable, more attractive, and more varied in design. However, most of the houses from this second generation of home building still stand, fifty and sixty years old now, some well kept and others not so well kept, but all jammed into those original tightly spaced long lines that twist up the Wright Fork hollow through Neon, Fleming, and McRoberts, preserving for these towns those mining-camp images so prevalent throughout Letcher County, Kentucky, and much of Appalachian coal country.

Neon, Kentucky, home of radio station WNKY-AM, was founded in 1913 as a trading center for the immediate region,[4] drawing business during the 1920s, 1930s, and 1940s, not only from McRoberts and Fleming, but also from other nearby Letcher County communities: Seco, Hemphill (officially known as Jackhorn), Kona, Millstone, Thornton, and Mayking.

Tradition explains the origin of Neon's name through a story about a black porter on the old L&N line. Apparently the train made its stops in this town in such a way that the passenger car's entrance lined up with a large tree stump, onto which boarders could step or "knee on" to reach the car's loading steps. All of this supposedly saved the porter the trouble of stepping from the car to put down a portable riser. As a signal to waiting passengers, during the train's approach, the porter was supposed to have cried out "K-n-e-e o-n," meaning that each boarder should mount the stump before stepping to the train. In *Kentucky Place Names,* Robert M. Rennick expresses some doubt about this explanation, and speculates that the name might have originated simply from the action of an early merchant to advertise his or her establishment with the first neon sign of the region.[5] Present residents of the hollow, however, apparently accept the "black porter" story, even though the name's prevailing pronunciation matches better the Rennick rationale.[6]

The industrial-boom years of World War I and World War II became the peak growth periods for this Neon/Fleming/McRoberts stretch of Letcher County, Kentucky, as American manufacturing sought to meet the military needs not only of this country but of other nations as well. The 1950s through the 1980s, however, witnessed a steady decline in the area's population and economic base. Neon slowly dropped from a population of 1,187 in 1940 to 1,055 in 1950, 766 in 1960, and 705 in 1970. Fleming experienced the same kind of downward spiral: 1,193 in 1940, 943 in 1950, 670 in 1960, and 473 in 1970.[7]

In 1978, as part of an effort to preserve and improve municipal services for this approximately two-and-one-half-mile section of the Wright Fork region, Neon and Fleming merged their city governments. McRoberts apparently was approached with the same deal, but declined.[8] This consolidation, plus at least one annexation that occurred during the seventies, produced the new town of Fleming-Neon and a 1980 population count of 1,195. At the time of this writing this consolidated community claims to be a municipality of 1,500, the population number published on a sign at the southern entrance to Neon; but the 1990 United States Census figures show Fleming-Neon to have only 759 residents, of which 720 are white, 29 are black, and 10 are American Indian.[9] Thus this community apparently suffered a 36.5 percent depletion of its residents in just ten years.

By the 1940s Neon possessed the largest commercial district of any of the small towns between Jenkins and Whitesburg, the latter being the county seat of Letcher County. Such establishments as Wright Motor Company, providing sales and service for Chryslers and Plymouths; Taylor's Furniture Company and Taylor's Food Market; Meade's Service Station and Ed's Taxi; the Craft Funeral Home; Neon Dry Goods Store and the Jackson House Furnishing Company; American Dry Cleaners, the Cumberland Hotel, and the Neon Firestone Store; Simon's Hay, Feed, and Grain Store; the Neon Fruit Market, the Abdoo-McKinney Jewelry Store, and Dawahare's Department Store, among several other such businesses, provided the goods and services needed by coal-mining families of the region.[10]

Today all of these commercial concerns have long since closed, with the tall, rusting, neon sign of the now empty Dawahare's Department Store rising as a monument to this almost-dead town. As is the case for a vast majority of storefronts on Main Street, Neon, Kentucky, Dawahare's spacious plate-glass show windows are covered from the inside by muslin curtains, and the empty second-story windows of this large brick building look down on a business thoroughfare largely devoid of pedestrian or vehicular traffic. Within the three blocks of what is now downtown Neon there are only a half-dozen or so business establishments still in operation, with First Security Bank and Super 10 ("nothing over $10") appearing to be the healthiest. At the time of this writing, Neon, Kentucky, is not a ghost town, but it may be on the way to becoming one.

WNKY
Neon, Kentucky

The studios of WNKY-AM share—with the local Masonic lodge—the second floor of one of these otherwise closed business facilities. This station first went on the air August 31, 1956, with five kilowatts of power, and has since supplied its Letcher County listening audience with a mixture of country-and-western and gospel music, and with a Sunday format of religious programming, including several in-studio live broadcasts.

These broadcasts originate from the one and only production studio the station has, a sixteen-by-twenty-four-foot space that windows off the main control room. The furnishings of this production studio are simple enough: a four-by-eight-foot wooden table; a mike stand sitting on the floor at one end of this table, equipped with an adjustable swinging boom that allows the microphone to be positioned, high or low, on any point of the boom's arc; several metal folding chairs, used by musicians, singers, and visitors; and one high stool, for those preachers who do not want to stand but who find a degree of elevation conducive to their preaching.[11]

Out of character with the prototype of such facilities, there are no photographs or posters of gospel singing groups adorning the walls of this studio, and—unlike the production space discussed in the next chapter—there are no religious symbols, crosses or otherwise, on the walls. In these senses, therefore, this production studio is a neutral zone, but the Sunday broadcasts that originate from this space are far from neutral. They are prime examples of the airwaves-of-Zion genre of radio broadcasting, and WNKY appears less likely to make any sudden movement away from its airwaves-of-Zion programming base than either of the other three stations treated in these case studies.

All of the radio preachers I observed at WNKY utilize the station's production studio by standing or sitting at the table, facing the control room, and thus directing their sermons toward the main console, perhaps with the sense that one should face these broadcast controls just as one should face a live audience. That swinging boom of the mike stand would permit any of a wide range of speaker orientations, but these exhorters seem to prefer a body focus that directs messages pointedly at the perceived outlet to the airwaves of Zion.

The room contains another window that looks out on the station's entrance foyer. Nonperforming friends of the preachers or singers often stand at this window observing the respective production as it takes place inside. During one of WNKY's live religious programs, "The Words of Love Broadcast," this foyer is packed with people, several pressed against this window, empathetically involved with whatever is transpiring in the studio.

I first visited WNKY on May 21, 1989, when I was trying to identify the specific stations and live religious broadcasts to be used as the case studies for this volume. At that time, the morning programming was monitored and controlled by Wiley Vanover, and the afternoon programming by Teddy Kiser. These two were still at the main studio console when I returned to the station for visits in 1990 and 1991.

Vanover, in his late forties, told me during that first visit that he served as a supervisor in a Terry's potato-chip factory in Hazard, Kentucky, and worked at WNKY on weekends. With no official training in broadcasting, Vanover began his work with WNKY in 1974, first simply announcing high-school basketball games and then assuming his Sunday-morning role.[12] This is a man who obviously enjoys sitting at the console of this radio station, directing the flow of sounds that wing their way along the airwaves of this part of Letcher County. He is natural, relaxed, unhurried, "front-porch-friendly" (by his own words), and regional in speech—a male counterpart to Mrs. Hayworth of WMCT, Mountain City, Tennessee.

Operating as the Sunday-afternoon announcer, Teddy Kiser was only a high-school sophomore when I first encountered him in 1989, and he had been at his post less than a year. A teenager of slight stature and juvenile facial features, he looked even younger than he actually was; and I recall thinking that he was probably the most junior of the announcers I had encountered to that point. Nevertheless, he handled his job with confidence, and he appeared to be genuinely liked by the airwaves-of-Zion performers who come to the station on Sunday afternoons, there to send forth their evangelical sounds to a waiting Letcher County.

When I returned to the station for my May 26, 1991, visit, Teddy

Wiley Vanover at the WNKY console.

Kiser had just graduated from high school and was making plans to attend Alice Lloyd College in Pippa Passes, Kentucky. His goal was to become a secondary teacher. "Maybe history," he said.[13]

These individuals represent well two categories of persons I have found operating the main-studio controls during my field investigations of Appalachian Sunday morning and afternoon broadcasting: high-school students (males and females) and older workers (retired or second-job people), neither group having formal training other than in the basic operations of the studio console.

Vanover seems particularly well suited for the role he plays, Sunday-morning spokesperson for WNKY. He is a stocky man, perhaps five feet eleven inches tall, friendly and open, who addresses his WNKY audience in a warmly provincial idiom, rushing nothing (verbal expression or action), and assuming conversational tones that make Letcher County listeners welcome his voice into their homes.

Teddy Kiser's image has been somewhat different, but apparently it is equally appealing to station WNKY listeners. He has been, according to Vanover, a high-school student whom older people of the community liked—first because he is of the area, next because he is from a respected family, then because he is intelligent and works hard at his education (almost a straight-A student), and finally because he treats all station callers (seeking information or making requests) with respect. "He doesn't get on the air and show off," said one middle-aged female resident of Neon, whom I stopped and questioned on the street in front of the station.[14]

When I interviewed Teddy Kiser in 1989, he informed me that his father worked as a coal lab technician. His mother was a housewife who belonged to a local gospel group, the Good News Singers, who had just cut their first recording.[15]

In May 1989, WNKY's Sunday programming included only three in-studio live religious broadcasts and one remote from a Fleming mainline church. The in-studio programs included "Keep on the Praying Ground," 1:30–2:00 P.M., conducted by Brother Millard Bates, Jr., pastor of a Hemphill Freewill Baptist congregation, assisted by two singers and musicians; "The Brother James H. Kelly Program," 2:00–2:30 P.M., featuring the solo performance of Brother Kelly, a seventy-five-year-old Pentecostal preacher who functioned, however, as the assistant pastor of Corinth (Independent) Baptist Church in Fleming; and "The Words of Love Broadcast," 4:00–5:00 P.M., produced by Brother Dean Fields and the congregation of Thornton Freewill Bap-

Teddy Kiser at the WNKY controls.

tist Church, Thornton, Kentucky, the program and group this chapter features.

The first of these preachers, Brother Millard Bates, Jr., of Whitesburg, Kentucky, carries a surname prevalent in the Wright Fork and Boone Fork region. John Bates, "an Irishman from North Carolina,"[16] had been among the earliest settlers in the area, arriving in the North Fork vicinity sometime in the first decade of the 1800s and later amassing a considerable holding in timber. His descendants, like those of Joel Wright, still populate some Letcher County communities.

Brother Bates shares his thirty-minute program with two other preachers: Brother Arnold Noles of Neon, and Brother Earl Breeding of Seco. On this occasion in May of 1989, Noles did most of the preaching, filling the small production studio with a rhythmically chanted sermon that reminded me of the first delivery style I described in *Giving Glory to God in Appalachia,* punching the end of each linear segment of his rhetoric with that pronounced "haah" I have spent so much time describing.[17]

An impassioned speaker, Noles seemed to lose himself in his exhortation. Standing in one spot throughout his sermon, he placed his hands on the long studio table, leaned low over the microphone, and concentrated exclusively on his radio audience, whoever they might have been. In the process he appeared oblivious of me, of Teddy Kiser in the adjoining studio, and of the preacher for the next thirty minutes, who entered the room midway into Noles's sermon.

That next preacher was Brother James H. Kelly, one of the most interesting of the many Appalachian radio exhorters I encountered during this four-year study. This aging Pentecostal—in his mid-seventies, slightly stooped, but apparently healthy—comes to WNKY with his wife each Sunday afternoon at a little before two o'clock, wearing a denim work shirt (cuffs unbuttoned but not turned back), dark-colored cotton duck trousers (the top button undone to provide him more comfort at the waist), and broad suspenders (red or blue). When his time at the mike arrives, he places a well-worn Bible on the studio table, sets a small plastic jar of some orange drink beside the Bible, pulls that high stool up to the speaker's position, removes his shoes (dark leather high-tops with heavy laces), leans against the stool without sitting completely on it, adjusts the mike so that it is approximately eighteen inches from his face (giving him room to move his hands without constantly striking the instrument), shuts his eyes, brings both hands up to shoulder level, extends the index

The preaching of Brother James Kelly.

finger of each hand, and—slightly rocking to the cadence of his own voice—delivers thirty minutes of fast-paced rhythmical rhetoric that typically has no theme, but wanders disjointedly over numerous warnings, admonitions, pleadings, and imprecations, all "to the glory of God." Like the ancient biblical prophets he emulates, he pleads with the "children of Zion."

"My friends, God loves you. I know he loves you," he shouts, his face flooded with concern, his index-finger pose momentarily discontinued in favor of open hands, thrust forward, palms up. Then suddenly, with equal force, jaws tense, eyes still closed but facial muscles communicating anger and finality of judgment, now back to emphasizing his charges with jabs from those index fingers—two kinetic forces pointed directly at the microphone, thrown up or out at appropriate moments—he tells his listeners what God will do to them if they do not get their lives in order, if they do not seek "salvation": "The lake of fire, Friends! The lake of fire!"[18]

Kelly had "some singers years ago," he said, but "for the last twenty years" he has been "all by himself." His wife comes to the studio with him, but she only sits in the foyer and listens, not participating in any active way.

This woman is a faithful supporter of her husband, however, and Kelly volunteered that she and he had read the Bible aloud to each other—cover to cover—a total of four times. In 1989 they were on their fifth reading, a process that must be very slow, given the struggling way Kelly reads Scriptures, frequently adjusting his reading glasses and the distance of the Bible from his eyes, providing his own peculiar pronunciations to Old and New Testament characters and places.

"I'm Pentecostal," he noted, "but I preach at Corinth Baptist Church. That don't matter," he added. "They get the Spirit."[19]

Brother Dean Fields and the Cast of "The Words of Love Broadcast"

When I first witnessed WNKY's Sunday programming, I was also impressed by Brother Dean Fields and "The Words of Love Broadcast," primarily because Fields involved a substantial part of the active membership of the Thornton Freewill Baptist Church. That afternoon there were almost forty people—men, women, and children—crowded into the production studio, the adjoining foyer, and a receptionist's space directly off this foyer, to sing, shout, and praise, blending their religious expressions with a wide range of traditional components of a congregational confraternity. As I mentioned in chapter 1, this is a church that transports itself to a radio station—almost in its entirety—to share its worship and fellowship with a wider audience, to follow the dream of the church's pastor, Brother Dean Fields, and to support each other in their ongoing struggles for personal stability and spiritual fulfillment, with this latter purpose apparently being the most important.

At the time of this writing, Brother Dean Fields is a forty-eight-year-old illiterate; called to a free grace, premillennialist, impassioned, markedly nontheological, "call upon Jesus" gospel; heavily influenced by the Pentecostal principles of faith healing, the gift of tongues, and "signs and wonders"; and possessive of a certain charismatic power over his small Thornton Freewill Baptist congregation. His focus is always evangelistic rather than instructional, and he colors his exhortations with a celebratory spirit that occasionally has his congregation laughing with joy.

A burly man, not overly tall (perhaps five feet ten and a half inches tall), but broad shouldered and husky, Fields possesses a high energy level that conceals the shakiness of his health. During what he says were twelve operations in 1981 and 1982, he had a sizable portion of his lower intestinal track removed to stop the spread of a virulent cancer, and today he suffers from frequent illnesses that he has not fully characterized to me.[20]

Fields founded his Thornton Freewill Baptist Church on June 6, 1982, soon after he recovered from his last stomach surgery. He had been involved with preaching prior to that time and had a radio program, but apparently his service to this calling had been spotty and unsteady before this traumatic series of experiences under a surgeon's knife. In addition, prior to the early 1980s Fields had suffered several bouts with alcoholism, times when he "fell away" from his preaching.

Brother Dean Fields now speaks of this series of operations and his liberation from alcohol as his period of rededication and as the beginnings of his work with the particular body of people who now constitute Thornton Freewill Baptist Church and "The Voice of Love Broadcast" group. In a very large sense, he depends as much on these people as they depend on him. Both this preacher and his congregation lean on each other for stability and purpose, in the process finding escape from problems they individually might not prove capable of handling.

According to his often-repeated account of the beginnings of this church, Fields borrowed a hundred dollars to get Thornton Freewill Baptist started. This was during the first week of June 1982. His congregation, a small force of followers, several of whom he says—with them nodding their heads in agreement—he "pulled from the gutter," began their meetings in an unadorned wood-frame, roof-sagging structure that still sits beside Thornton Creek, a building inherited from some earlier group's effort to establish a church.

Two years later the congregation obtained an old house trailer, gutted it, and set it up at the rear of the original facility, using this second structure as a church kitchen and fellowship hall. Finally the congregation built their own sanctuary, a sturdy, boxy and unadorned and unpainted, concrete block structure that sits sufficiently high above the valley floor to avoid the frequent flooding of Thornton Creek. A newspaper clipping pinned to an entryway bulletin board in the new building shows the membership's first church standing in several feet of swift-flowing flood water, an event that occurred on May 7, 1984.

The Thornton, Kentucky, community of homes and other structures is clustered along the lower reaches of Thornton Creek just above and right at its juncture with Bottom Fork (also known as Mayking Creek), another tributary of the North Fork of Kentucky River. This is an unincorporated community, and the only structure, other than a couple of churches, that bears the name "Thornton" is the minuscule post office that sits near the confluence of Thornton Creek and Bottom Fork.

Coal mines that once gave employment to hundreds of Thornton, Mayking, and Millstone families were opened during those boom years immediately prior to and during World War I; but, like the mines of the Neon-Fleming-McRoberts stretch, the Thornton/ Millstone vicinity ones have been closed for several decades, leaving only a number of derelict facilities—crumbling, overgrown, rising from residual piles of slate and other unearthed mineral matter, and all of these sites linked together by now-trackless railroad beds, themselves becoming reclaimed by the aggressive growth of rich Appalachian vegetation. Some of these ruins still bear the markings of the South East Coal Company, the corporation that developed this immediate area of Letcher County, and which left its initials as the name of one nearby town, Seco,[21] but which also left a more indifferent message in the scarred and debris-strewn hills of this Kentucky region. When the coal deposits were depleted, what was valuable and could be carried out (the railroad tracks, rolling stock, and reusable machinery) was removed; that which was expendable (these decaying buildings—and the people) was left.

Some three hundred yards "up the holler" from Thornton Freewill Baptist Church are the remains of one South East Coal Company operation, wood and steel structures that once constituted the tipple of a mine. Today these facilities are rusting, rotting, and unsightly, creating a danger for children who might play among the ruins, and a steady creek-polluting run-off. One of the blights afflicting many regions of eastern Kentucky, southern West Virginia, and southwestern Virginia is that prototypic abandoned coal-mining facility, a collapsing network of slope-hugging structures, resting on embankments of stone ripped from the bowels of the respective hill, covered with those layers of slate and unusable coal chips, surrounded by those scatterings of rusting equipment companies found less costly to abandon than to remove, all of this constituting a monument to the environmental irresponsibilities of an earlier (and sometimes present) age.

To reach Brother Dean Fields's church, one would exit Highway 119 at Mayking (between Jenkins and Whitesburg), travel northeast toward Neon, cross Bottom Fork at the Thornton post office, travel up the east side of Thornton Creek approximately one half mile, cross Thornton Creek, and then double back in front of the crumbling mine tipple mentioned above, following an unpaved, one-lane driveway that once was the bed for the railroad track. The church sits in the creek's flood plain, but a four-foot-high ridge of stone and dirt keeps the water in the creek bed during all but the heaviest of rains.

In 1990 Thornton Freewill Church claimed a membership of 173 "saved souls," approximately 100 of whom were active and regular church attendees, and "all"—according to church member Geraldine Boggs—"brought to the Lord" by Brother Dean Fields.[22] Boggs and other members of the congregation—like seventy-eight-year-old Martha Adams—claim that Brother Fields has been the main instrument in their "redemption," each telling his or her story of how he helped them "pray through" to "a saving grace."

"I've never been to a better church," says Adams. "It's the love that's here. And I practice tongues. Some preachers don't like that no more. Brother Fields said, 'You speak whenever the Spirit calls and whatever the Spirit calls.'"[23]

Labeling himself a "pastor" and not a "master," Fields has gathered this flock from the various small communities already mentioned in this chapter, Thornton, Mayking, Millstone, Seco, Fleming-

Neon, and McRoberts. One member joined Thornton Freewill Church and then moved completely out of the region—to Abingdon, Virginia; but this individual, one of the preachers the church ordained, continues at the time of this writing to attend the Thornton services three times a week. Another member actually lives in Lexington, Kentucky, but maintains her formal membership with the Thornton fellowship, returning faithfully to the region for extended spring and summer stays. Fields himself lives in Jackhorn (Hemphill), northeast of Thornton.

Dean Fields's formal education took him only through the third grade and provided him, as I have already established, absolutely no reading skills. He talks of learning his ABCs when he was forty. Operating under such an educational deficit, a preacher can survive only by a heavy memorization of Scriptures and/or through the assistance of readers. Fields has taken both options and in the process become a prototype of that genre of Appalachian preacher who manages to exhort without being able to read the document from which he or she exhorts.

In his pulpit, and while on the radio, Fields functions by holding in memory a wide-ranging selection of Scriptures and by having one of the church's literate deacons read those Bible passages he has not committed to memory. In the latter circumstances, however, Fields needs to know the book, chapter, and verse of the desired Scriptural segment.

During sermons a partnership develops between Brother Fields and his reader of the moment, with the latter rendering the verses and Fields providing interpretation and application. The two contributions to such homiletic moments are so synchronized in spirit, tone, rhythm, and volume, that the reader becomes a part of Fields and Fields a part of the reader. The effect is not unlike the Ward-brothers coalition I witnessed at Sister Benfield's church, except that in that instance the connection was a line of mike cord and a kinetic partnership, rather than the verbal union I am now describing.

During my two decades of study of Appalachian religious traditions, I have witnessed other Scripture-reader/Scripture-expounder partnerships, one of the most colorful of which was a husband/wife team. In the early 1970s this couple had a program over WMCT, Mountain City, Tennessee, and the woman helped the man by interjecting into her husband's sermons additional scriptural material whenever he seemed about to run out of things to say. The new verse

Brother Dean Fields, *center,* and singers, Thornton Freewill Baptist Church.

of Scripture would always generate a fresh burst of exhortation, sometimes on a theme separate from the one the preacher had been pursuing.[24]

This, however, is not the technique employed by Brother Fields, who always maintains control over the direction of his sermons, asking his reader to get prepared with a new section of Scripture long before he (Fields) calls for its rendering. Nevertheless, if the passage is lengthy—several verses perhaps—Fields will not wait for its completion, but will expound on each verse, or even part of a verse, as it comes from the lips of the reader, often resulting in an overlapping set of sounds, each building on the other.

When I asked Fields why he did not, even at this late date in his life, enter a literacy program, he said simply that there were other "callings" placing more urgent demands upon his time and energy, an allusion to his sense of mission in building the Thornton church. I wondered, however, even while asking this question, if the acquisition of such additional stills would not unsettle his pulpit style, requiring him to spend years developing a new set of procedural dependencies.

Brother Dean Fields believes that the establishment of his church in the Thornton Creek hollow has changed the nature of life in that immediate region of Letcher County. "Things has gotten a lot better along this creek since we moved in here," he says. "There was a lot of mean folk along this creek, doing a lot of mean things. But they respect this church. The door's always unlocked, day and night, and the only thing I've lost in nine years was two light bulbs, and they brung them back, with a note saying they was sorry."[25]

When Fields tells stories such as this one about the light bulbs he can be very convincing. His unbridled expressions of "love" (and that is his word), for his church, the people of Thornton Creek, his radio audiences, and mankind in general, lend credibility to his claims for kindnesses shown him; but his intense rhetoric can be unsettling to anyone not accustomed to such outpourings of sentiment. Indeed, my own first encounter with this man—at WNKY—left me somewhat flushed: he wanted to tell me why he had named his program "The Words of Love Broadcast," and he spoke with a level of ardor I did not expect in an opening conversation. The folklore surrounding the exploits of "Bad John" Wright depicts that Letcher County native as a man of very few and very controlled words. Fields certainly did not fit that stereotype.

Services at Thornton Freewill Baptist Church

I visited Thornton Freewill Baptist on three occasions, two Saturday-evening services and one Sunday-morning service. I found the Saturday evening meetings more intriguing than the Sunday gatherings, in large part because they were such unstructured happenings, relying heavily upon the spontaneous contributions of the congregation, and imbued as they were with such intense levels of emotionalism. Like Sister Kathy Benfield's church in Burke County, North Carolina, the Thornton congregation does not rely solely upon the contributions of the preacher or preachers: They themselves become the service. "You come here tonight to help me praise the Lord," Fields said to his congregation on Saturday, June 16, 1990, "so let's get to praising."

The interior of Brother Dean Fields's church shows more attention to decoration than the exterior. While the unpainted concrete blocks of the outside communicate a Spartan barrenness devoid of any passion for embellishment through color or design, the inside's busy conglomeration of inexpensive religious plaques, pictures, tapestries, and other artifacts suggests—like the situation in Brother Kittinger's church—a taste for the tinselly and the tawdry, devoutly cherished although each of these faith symbols have become.

A large homemade cross—structured so that all four arms of the icon become recessed compartments, each wired with small lights and covered with glass—dominates the wall behind the pulpit. Then around the full circumference of the sanctuary are small whatnot shelves, attached to the walls, each holding a kerosene lamp. During night services, Fields frequently intensifies the mood of the meeting by lighting the lamps and the cross, while turning off all overhead units, procedures that invariably unloose the passions of his congregation.

This interior is inexpensively carpeted, and all four walls are covered by paneling that also appears to have been economy-priced. Regular illumination is afforded by yet-to-be-hooded fluorescent ceiling panels, and seating is provided by a mixed set of wooden pews, one group coming from the old church and looking homemade, and the other group seeming newer and more professionally constructed. At the front of the sanctuary there is that traditional section of raised flooring, upon which rests Brother Fields's pulpit, four or five standing microphones, a couple of portable amplifiers for musical instruments, and a horseshoe of benches for the musicians, preachers, and deacons. On the front lip of the pulpit there usually is a neat line of Styrofoam cups, each filled with water for consumption during services.

All of the physical features of this meeting hall combine to create a very functional and flexible worship space, with emphasis placed upon providing room to move about. One factor contributing to this functionality and spaciousness is the large open section between the pulpit and the first row of pews (the pit), an area that accommodates the frequent down-front gatherings of the worshipers—to kneel and pray, to stand and shout, or to cluster for laying-on-of-hands circles of blessings for one or more individuals.

The openness of this part of the meeting house also allows Brother Fields to bring his sermons off the platform and directly to the congregation, moving—mike in hand—deeply into the audience zone— shouting, jumping, waving his arms, embracing members of his church, stamping his feet during rapid-fire bits of rhetoric, and just generally playing out to the fullest his shirt-sleeved, tie-loosened, work-up-a-sweat delivery style. Indeed, it is in this area of the church that Fields spends most of his time while preaching, frequently charging up the center aisle to take his message to some particularly responsive cluster of auditors.

In Pentecostal churches the pit is often the scene of "laying on of hands" and "swooning in the Spirit" sessions that culminate in several worshipers lying supine, arms frozen in the particular posture each individual held at the time of being eased backward to the floor. Such a scene will be examined in chapter 5. During the three services I attended at Thornton Freewill Baptist Church, however, no "laying on of hands" episode produced this result.

As was true in the churches of Sister Benfield and Brother James Kittinger, amplified sound is critical to the worship style of the Thornton Freewill congregation. The volume of instrumental music, singing, preaching, and testifying is intensified to such a degree that there are times when individual sounds cannot be distinguished from all the other noises, thus blending everything into one unified expression. In addition, the exuberance of individual singers and preachers is heavily dependent upon these high-volume dynamics, to such an extent that the expresser's passion diminishes when sound levels are suddenly dropped. I watched Brother Dean Fields lose a considerable degree of his concentration, motivation, and ardor when the mike into which he was preaching suddenly went dead, the result of a short in one of the connections. Although he was exhorting with an unassisted

131

A testimonial letter is read at Thornton Freewill Baptist Church.

volume sufficient to be heard in every corner of the church, and probably on the outside, he reacted to this loss of electronically amplified sound as if steam had been withdrawn from his spiritual calliope, experiencing some reduction in fluency before quickly shifting to a functioning mike. A deacon immediately corrected the problem in the faulty wiring, and the mike was returned to use just in time to be taken over by one of the singers.

The church's Saturday-evening services begin informally with singing, testimonials, requests for special prayers for troubled souls, infirm church members or family loved ones, and problems in the community, state, or nation. There appears to be no predetermined order or format, but one thing Brother Fields does early in a service is call the entire congregation to the open space in front of the pulpit, there to warmly embrace each other and perhaps to sway from side to side with arms raised, singing, praying, or shouting. Such initial congregational encounters make any beginning periods of reserve short-lived, with the crowd conditioned for early and rapid accelerations in joyous enthusiasm.

This is a heavily tactile church—hugging, laying on of hands, handshaking, and the like; and each service is filled with a number of scenes in which several worshipers gather around one other communicant, there to offer support, comfort, or "healing" through touch and a communal consciousness. During the first Saturday-night service I observed (June 16, 1990), Brother Fields called one aging man to the open space, announcing that the individual had been with the church two years and that it was about time all the other members showed their appreciation of him by coming forward and hugging his neck; which they did, lining up so that each could have her or his turn at this communion.

Moments such as this become episodes of intense emotionality, with the recipient of the touching either bent low and yielding or standing tall in exaltation, arms thrust joyously into the air. On that Saturday night this old man cried, genuinely touched by the tribute.

Formal preaching usually does not become the main focus of these Saturday-evening services. Instead, emphasis is placed upon singing (solo, group, and congregational), shorter testimonial-like exhortations provided by a number of the preachers, spontaneous expressions from individual church members, periods of shouting and crying, requests for those laying-on-of-hands blessings and healings, and such other forms of impassioned expression as might develop.

When I observed the church on May 25, 1991, it was my fourth exposure to most of the members of this congregation (either at the church or at the radio station), and the group had become comfortable with my presence. As a result, the service that evening was a particularly exuberant one, providing examples of a wide range of this fellowship's worship behaviors.

About midway into the service one of the church's six preachers began to exhort, and I moved toward the front of the church to take pictures of him. The congregation was relaxed enough with me that they tolerated my wandering at will around the sanctuary, snapping pictures whenever I wished. During these experiences, I saw evidence that my camera never seemed to depress the emotionality, but instead became another stimulant in the total process.

As I crouched to take my photograph of the preacher in question, I heard a female member of the congregation begin to shout. At this moment I also became aware of a rush of footsteps immediately behind me. Somewhat startled, I turned to discover that a male worshiper had begun to "run in the Spirit," charging around the sanctuary in a rapid sprint, engaging in no shouting, but completely captured by some intense motivation that showed graphically in his face as he ran.

This man had been overcome by an impulse to translate emotions into action, "possessed," he later argued, "by the Holy Spirit." Furthermore, in this act of running he not only had provided his own nervous system an outlet for release, but he had generated in other members of the audience tears, laughter, shouts, cheers, and explosive bursts of approving applause. One woman began to jump up and down, clapping her hands continuously. Another female was laughing and crying at the same time.

At the back of the church a small boy had been playing under a pew with a set of plastic soldiers, but suddenly he was standing and attentive to all the actions of the aroused adults around him. He stared at them and at the sprinting figure, casually surveying the scene at the same time that he maneuvered a toy military vehicle along the back of the pew in front of him. Then he momentarily stopped that last action, as he focused—nonjudgmentally—on the running figure.

Brother Fields quickly joined the run, but he lasted only two of those fast-paced laps before stopping to lean breathlessly against a front pew, rejoicing in what to him was a perfectly legitimate form of "sanctified" religious expression. One aging woman continued for some time to clap her hands, raising them above her head as she did so. A young mother and her daughter sat on the left front pew and

Practicing a gift: "Running in the Spirit."

laughed throughout the episode, but approvingly so. Indeed, all congregational reactions suggested complete acceptance of the behavior.

The runner continued his rigorous expression of religiosity for six or seven laps around the church's interior before exhausting the initial burst of energy ("Spirit") that had overtaken him. Finally he returned to his pew, receiving another round of applause as he did so. However, he did not immediately sit, opting instead to stand in a posture of quiet repose, arms raised, chin lifted, and eyes shut. The only motion I noticed in him was a slight side-to-side sway.

One interesting thing about the runner was that prior to this happening he had appeared to be a relatively uninvolved member of the congregation, shy and reserved in his behavior, sitting quietly in one of the back pews. He had not moved about the meeting space the way so many other worshipers had done, rushing to particular pockets of fervor; and he had not shouted or clapped his hands in response to preaching or testimonies. Still the emotions of the evening must have been working on him, tensing him for this one explosive moment of expression.

I learned later that "running in the Spirit" had been this man's one form of response to the emotive displays that constantly developed in the services of Thornton Freewill Baptist Church, his "gift." Apparently he did not do it every time he came to church, but his "Spirit exercise" was regular enough to create a congregational expectation of the behavior. He was not a singer or a testifier, but he could enliven a worship service with his sudden charges around the sanctuary, bringing people to their feet and making them feel happy. He ran not away from something, but after something; and his fellow revelers shared the joy of the chase, whatever it was he pursued.

Tension had built and been released, passion channeled into this compulsive burst of kinetic energy, then leaving the communicant spent, but tranquil and pacific. He soon stopped his swaying and stood almost perfectly still, trancelike, with one woman staring empathetically at him, caught in some experience of her own. The preacher had ceased his exhortation, and for a brief moment quietness settled over the congregation, providing some relaxation before the next round of frenetic worship began.

During my various observations of Appalachian religious practices—particularly among Pentecostal-influenced traditions—I had seen "dancing in the Spirit," "swooning in the Spirit," "the jerks," and numerous individual acts of "practicing tongues"; but this was the first episode of "running in the Spirit" I had witnessed. The exercise belongs to a well-documented and frequently examined genre of psychophysical religious phenomena.

For example, literature descriptive of events of the "great western revival" that swept the settled regions of the Cumberland at the very beginning of the nineteenth century are replete with accounts of "revival exercises," extreme physical behaviors attributed either to disturbances of mind present in the "sinner" or to releases relevant to "redemption": jerking, running, dancing, barking, falling, and rolling, among others. The exercise I witnessed at Thornton Freewill Baptist Church fell within this family of practices, although it varied sharply from the frantic run-in-place behavior that some witnesses have described.[26]

Later in the service a second event occurred that seems worthy of mention. Sister Dorothy Dawahare, sitting one pew in front of the man who did the running, suddenly collapsed forward, crying uncontrollably, her head on her knees and her shoulders racked by a spasm of jerks. Immediately a circle of ten worshipers formed around her; bending over her; hands reaching out to touch her head, shoulders, or back; each participant praying, crying, or shouting; a chorus of emotional expressions concentrated on this one troubled individual.

My first reaction was to interpret this happening as a scene being played out around a "sinner," one seeking to "pray through to conviction," a woman gripped by some perception of guilt, shame, or personal imperfection. I expected, any moment, to see the woman led forward to that open space before the pulpit, there to sink to her knees and seek "redemption" through a tearful confession of "sins" and a joyful proclamation of "call to salvation." It also flashed through my mind that on the next day, Sunday, I would witness her baptism in Bottom Fork, three or four preachers to each side of her, plunging her back-first into the cold water of that stream.

I had observed just such a baptism on June 17, 1990, when I visited this church for one of its Sunday-morning services. That afternoon I witnessed a creek gathering as exuberant as any of the score or more I have observed during the last twenty years, and frankly I was eager to view a second such happening.

My initial interpretation of Sister Dorothy Dawahare's problem, however, was not correct. This woman had recently lost a twenty-one-year-old son, a consequence of a one-car, single-passenger accident in which excessive speed had been the cause. Sister Dorothy, however, had not viewed this happening as an act of man but as an act of God, one of the directives of "Providence." God had "called" her son "home,"

Church members clustered around Sister Dorothy.

she said. The young man, she told me later, had not left this world of his own accord or completely by his own doing. In fact, "he hadn't been ready to go" and perhaps not "prepared to go," she cried. "Why," she sobbed, referring to the entire set of circumstances that took her son's life, "did this happen?"

Her question is one asked by many predestinarians when personal tragedy strikes. If God controls all, why does he appear to be so indifferent to consequences? At the time Sister Dorothy Dawahare collapsed in her pew of the Thornton Freewill Church, she had not found an answer that "added up" for her, "comforted" her, she said, an answer she could tell herself and all around her with a pride in comprehending, in knowing.

"I don't understand," she cried to those around her. "There are old and sick people who want to go, who need to go and can't. My son didn't want to go."

"That don't really matter," advised the woman sitting beside her, both hands resting on the sister's right shoulder. "Besides, man is man, and God is God."[27]

This last statement went unexplained and appeared to be the most detailed advice Sister Dorothy Dawahare received that evening from her fellow members of Thornton Freewill Baptist Church. Eventually these words of advice and the laying-on-of-hands she received from her friends and fellow believers brought her some tranquillity. Her crying subsided, and like the man who had run his laps around the sanctuary, she found her own quietness, now permitting the woman who had provided the "God is God" commentary to hold her gently.

The scene had occurred near the end of the ninety-minute service, and Brother Dean Fields allowed the event to end the evening's flow of emotions, not as a downer, however, but as a positive expression of fellowship and support. It seemed obvious that Brother Fields viewed the woman's explosive emotions as fruitful, a working out of something, and not just a futile cry for meaning. He did not hear the "God is God" line, but it probably would have sufficed for him. Furthermore, he would have pronounced this entire scenario as "good," based on the "love" and support this Sister received from the congregation.

Brother Dean Fields's Homiletic Message

"Love" is an important word for Brother Dean Fields. He uses it often, employing it always in his verbalizations of the meanings of his church and his radio broadcast: "We love each other," he says, speak-

The baptism in Bottom Fork.

ing of his cadre of thirty or forty faithful followers. "That's what it's all about," a sentiment repeated by his wife and three grown daughters, all devout and participating members of Fields's group.[28] He also has a grown son, but apparently views him as "fallen away."[29]

Fields's religious philosophy is largely positive, avoiding hellfire-and-damnation pronouncements, like Brother Kelly's "lake of fire" metaphor. He does, however, occasionally attack principles, behaviors, or institutions he believes to be both religiously and socially wrong: drugs, the lottery, lack of prayer in public schools, all restrictions of what he considers legitimate religious expression (municipal actions to circumscribe street preaching and similar types of religious expression), liberated sexual practices (particularly homosexuality), divorce, abortion, and drinking. All of these practices, or prohibitions, he says, work against the "will of God" and the "good of man."[30]

Alcohol receives a great deal of Brother Fields's homiletic attention, his comments in this area frequently centering around some discussion of his own earlier drinking problems. He tells both his church and his radio audiences about a period in his life when he would "come home drunk, scaring the children." However, he likes to follow this confession with his story about one of his daughters entering high school, "without a new dress to wear," but "happy" because "her Daddy" had recently stopped drinking.[31]

The first time I heard this narrative was during Fields's June 17, 1990, Sunday-morning sermon. It was Father's Day, and Fields had begun the service that morning with all the children coming forward to present to their fathers crayon-colored paper cutouts of a fish with the child's Father's Day message on the back side, messages the children would read aloud to the congregation. These cards were the products of the morning's Sunday-school lesson.

For one child there was no "Father," requiring another of the "Dads" to fill in as surrogate, a pretense that both child and man seemed to understand. Apparently it made more sense to all concerned that the card be presented to a male figure, rather than to the mother in question.

Following this scene, one of Fields's daughters tearfully sang, "Oh, Mommy, Why Did God Take My Daddy?" but with no precise application being made of this song to the fatherless-child scene. Next Brother Dean Fields preached his sermon; and when he got to the story about his drinking, his frightened children, and his daughter going to her first day of high school without a new dress, but happy,

141

The children at Thornton Freewill Baptist Church with Father's Day tributes.

every woman in the congregation was crying—as were several of the men—as was Brother Dean Fields.

The entire episode could have been described as "maudlin," but it also could have been characterized as "real," "sincere," or "honest," an earnest expression of strongly felt sentiment, from both Fields and his church. I report the scene not only as an example of the emotional level of Fields's sermons but as representative of this pastor's handling of such events as Father's Day and such problems as alcoholism.

Social and political issues, however, are usually addressed with considerable simplicity. Whether the topic be deterioration of the American family, drugs, abortion, state lotteries, prayer in public schools, or alcohol, Brother Fields's recommended remedy will be "Turn to Jesus," thus leaving all details of the problem to the "workings of the Lord." That advice pleases his congregation, and apparently they make the principle work satisfactorily for themselves, most of them having their own private devil or devils with whom they have wrestled.

As was suggested earlier, when Brother Fields is preaching he is constantly in motion, usually operating with one of the several mikes that are attached to long cords, thus allowing him to move out to the congregation, pacing in the open area or surging up the middle aisle. At these times he is often met by one or more of the worshipers, rushing to embrace their pastor or seeking a laying-on-of-hands blessing. At no time do Thornton Freewill members feel compelled to remain in their pews. Instead, they assume a freedom to move to any place in the meeting hall they wish to visit—drawn to whatever center of ardent expression that might be developing, to the support of a Sister Dorothy Dawahare or to the presence of a Brother who likes to "run in the Spirit." The point here is that in the Thornton Freewill Baptist Church there is no tradition for remaining seated in one pew. When "the Spirit moves," so does the congregation. That's what made the runner's initial reserve so noticeable.

"The Words of Love Broadcast"

I mentioned before that the first factor attracting me to an examination of this program was the size of the cast of characters—a substantial portion of the entire Thornton Freewill Baptist membership, children as well as adults, so many people in fact that the production studio described earlier could hold no more than a quarter of them at any one time. In the typical "Words of Love" gathering there are chil-

143

dren (toddlers to nine- or ten-year-olds), teenage girls (but no teenage boys), young married couples (several pairs), a middle-aged group, and usually a couple of senior citizens, all of whom actively participate in the broadcast if they desire, but some of whom merely serve as spectators, often clustering at the window between the foyer and the production studio.

These second-floor, un-air-conditioned facilities can become very warm during spring and summer months, and the crush of thirty-five or forty bodies intensifies the problem. This was particularly true on May 26, 1991, during my second visit to this station. It had been a hotter-than-usual May Sunday in Letcher County, and by 3:45 P.M. WNKY's production studio and adjoining entrance foyer were already uncomfortably warm. By 4:30 the air was stifling. Still Brother Dean Fields and his church threw themselves into a broadcast that was about as vigorous as could be. The heat was especially oppressive in the production studio itself, where the shirt-sleeved musicians were soon dripping with perspiration. By the end of the broadcast Fields's white shirt was completely soaked, but he still hugged each singer or testifier after his or her contribution to the program.

This is a sixty-minute show that follows no fixed format. The time distribution, however, adheres roughly to the following formula: Thirty-five to forty minutes of hymn singing; five to ten minutes of prayer and personal testimonies (sometimes more for the latter); five to ten minutes of announcements, recognitions of financial supporters, and words of encouragement for infirm Thornton Freewill Church members or regular broadcast followers (considerable calling of names during this period); and ten to fifteen minutes of preaching, by Brother Fields or one of the other ordained church members.

Fields's general rule is that any church member may participate in the broadcast whenever he or she feels ready to do so—ready in Spirit and in will. Regardless of talent, if an individual wants to sing a hymn on "The Words of Love" program, that individual is joyfully allowed to sing; and if he or she wishes to throw in a testimonial, that's definitely encouraged, even tearfully so.

One small group of regular attendants, however, never formally contribute to the aired program, satisfied only to be present and supportive. According to Fields, other individuals come to the studio for a year or more before getting up the courage or the motivation (the Spirit) to participate in any formal way. My impression is that performances in the church become preparatory for eventual movements to the studio mike, and that involvements with the radio show constitute evidence of a more complete commitment to the church fellowship. In addition, family dynamics influence this entire process. For example, there appears to be considerable parental pride in a child's early performance on the broadcast: During the first "Words of Love" program I witnessed (May 21, 1989), one young mother brought her five-year-old daughter to the mike to sing a song the child had learned in Sunday school. The child sang so softly that I wondered how much of the hymn had gone out over WNKY, but Fields praised the contribution in words that made the mother proud and reinforced the child's courage to perform.

Regardless of all these individual contributions, the actual formal participants on any one broadcast will be limited to perhaps fifteen or eighteen persons (including the musicians), in part because of the limited space in the studio. The remaining individuals simply gather in somewhat the same way that they do in church, perhaps singing in the background and perhaps contributing to the shouting if that kind of response gets started, but otherwise just drinking in the atmosphere and congregational spirit of the broadcast.

It has been my observation, during visits to a host of airwaves-of-Zion broadcasts, that there is a category of in-studio spectator, particularly older people, who simply gravitate to these programs out of a lingering sense of the "romance of radio," a feeling about this media that may be long gone from the thinking of the average American. These are the people who, if there are several productions in a row, may sit through all of them. That was the case for three elderly women who were in the production studio of WLRV, Lebanon, Virginia, when I visited that station on March 26, 1989.

The young children present for "The Words of Love Broadcast" are especially fun to watch, having as they do a relatively free run of the facility, even to the extent of running into the main console room where Teddy Kiser sits. This room, however, is supposed to be off-limits to the children, and Kiser usually talks one of the older boys into standing guard at the door to this area, but just inside so that this child feels the importance of being where other children cannot enter.

No one seems to mind, however, the children's diversions when they extend into the production studio during the actual broadcast, occasionally creating noises that must go out over the air. During that May 21, 1989, program, one preschool girl entertained herself for the full hour by playing around or under that production studio table I mentioned earlier, seemingly oblivious to all the emotional displays of the adults in the room.

Children of the airwaves of Zion learn to recognize the differences between parental emotions for which they should be concerned and parental emotions that they can accept as nonthreatening. One interesting scene for study is the oft-repeated picture of a child carefully monitoring the religiously aroused state of a parent or grandparent, carefully measuring the event in terms of its portent.[32]

Nevertheless, for the most part the children of "The Words of Love Broadcast" are not involved with the in-studio activities of the adults. They entertain themselves in the entrance foyer or at the end of a hallway between the production facilities and an office that opens off the foyer. In this hallway the younger boys enjoy playing with toy cars, and sometimes children will spread their coloring books or comics in this space. Suffice it to say that these children make themselves at home in the radio station, while the adults go on with their celebratory broadcast, relatively unmindful of their young charges.

In all of these children-related situations, an environment identical to that of a regular service at Thornton Freewill Baptist Church is created. As suggested by my description of the "running in the Spirit" scene, children play under pews or in aisles during all but the most boisterous of their parents' singing, shouting, and praising. During one Thornton Freewill Baptist Church service I took my laptop computer into the sanctuary, hoping to record my observations in that fashion. However, the move became an unfortunate one, since the children crowded around me, thus—I feared—distracting their parents. In short order I placed my computer in its case and returned to recording handwritten field notes.

Fields has a half-dozen guitar players in his congregation, at least four of whom are regularly present during the radio broadcasts. This is the form of participation Fields uses, both in the broadcast and in his church, to involve young men. While discussing the scarcity of these young men in his church, Fields confided to me that he had discovered this one way of attracting the youthful male to Thornton Freewill Baptist services. There did seem to be a definite gender differential at play relative to the musicians and the singers, with the former being all male and the latter being all female.

The contributions of these singers (a participant category that appears to include every woman of the church) consume the major portion of the program's broadcast time, as is also the case in any Thornton Freewill church service. In solo performance or harmony groups, these women pour themselves into the emotions of the broadcast, of-

A child plays as adults perform.

ten crying as they sing. In fact, without these singers the program would lose much of its fervor.

Brother Fields's favorite on-the-air practice is to share the microphone with a singer, shouting exhortative comments between or over the verses of the hymn. The emotional displays of one of his regular female singers seem particularly dependent upon this process, and when these moments occur she frequently interrupts her own song with a number of shouts of exultation, each time shutting her eyes and throwing her arms in the air. These episodes then result in other women joining the shouting. On one occasion an especially intense shouting session developed in the studio, with the result that the foyer emptied of bystanders, everyone who could do so pushing himself or herself into this scene of high emotion.

What is perhaps most interesting to watch are the manifestations of Brother Dean Fields's passion for the microphone. As suggested by behaviors described above, he is never far from it, even when someone else is singing or testifying. On these occasions he closes in on the scene, speaking over a shoulder of the respective communicant or from the side, and interjects impassioned comments supportive of the singer's or testifier's message, working in concert with the tone, tempo, and sentiment being generated. On these occasions a definite reciprocity of enthusiasm emerges, not unlike the most intense emotional partnerships that develop during the paired performances of Fields and his Scripture readers. In addition, the singers and the testifiers never seem to object, choosing instead to feed off the passions of their pastor.

At the time of this writing, Brother Fields and his church pay WNKY two hundred dollars a month for this one hour a week of air time, a considerable sacrifice for this group of relatively indigent individuals. Nevertheless, no one affiliated with the program seems to question this expense, believing as each does that the broadcast "serves the Lord." In support of this belief, group members have their favorite stories about some impact the program had on a "sinner friend." Fields himself is convinced that each Sunday the broadcast reaches "hundreds" who "need Jesus." As verification for his claim, he tells of letters and phone calls he has received, each involving a story of "a sinner saved."

Airwaves-of-Zion listeners are encouraged to call during the broadcasts, to make requests for prayers or specific music, or to report responses to particular program episodes. To accommodate these calls, one church member stays at a phone in the office adjoining the en-

Brother Fields at the mike with one of the singers.

trance foyer, occasionally taking messages to Fields in the production studio. At other times Brother Fields is actually called to the phone, there to give counsel or to pray with a distraught listener. During those moments a singer or another preacher will take charge of the mike: "Brother Fields is praying with Sister So-and-So. Praise God!"

When I visited the program on May 26, 1991, an episode such as the one described above did transpire, a scene I mentioned in chapter 1. Late in the broadcast Fields was called to the phone, and what he heard on the other end of the line pleased him considerably, resulting in several loud shouts of "Praise God!" After several minutes of listening he returned to the mike in the production studio and reported to his audience that a particular Sister—naming her on the air—had called to report the "salvation" of her husband. According to the account Fields then provided, again over the air, the couple had been driving down a Letcher County road listening to the broadcast when the husband suddenly pulled to the shoulder, got out of the car, knelt by the rear bumper, and started praying. It was later that the wife phoned Fields to report the results of her husband's experience. Inspired by this apparent success, Fields then told his radio audience that they too could find such spiritual satisfaction if they only fell on their knees "in front of the radio."[33]

It was an exciting moment for those individuals standing in the foyer, because the "call-in" Sister and her husband were known to the church. One remark was made about how long this Brother had struggled with his "sinfulness." An important point of emphasis here is that the names of this woman and her husband went out over the air, Brother Fields's assumption being that both would be proud of all factors in the narrative.

Almost in every way this "Words of Love" broadcast is a happening of intimacy. The various participants expose their emotions to an invisible audience, and parts of that audience respond in kind, carrying on dialogues in which many persons would never participate over such a public medium. To facilitate such personal exchanges—such intimate disclosures of pain, sorrow, and guilt—certain assumptions must prevail on both ends of this airwaves-of-Zion performer-audience connection: (1) All men and women are equally weak, a flaw in the very nature of the human condition; (2) no person can boast for having avoided the particular weakness that inflicts another human, the "there but for the grace of God go I" principle; and (3) there is no shame in having been in "sin," only in remaining in sin. Armed with these rationales, Brother Dean Fields can experience full identification with the "lost" individuals he seeks to help, and they in turn can find that identification with him. The dynamics of this reality produce a degree of trust that is often missing in less-intimate mainline religious groups.

A Closing Observation

It is difficult to ignore the interdependency that has developed between Brother Dean Fields and his congregation of followers: He has helped them find "conviction," and they apparently have helped him stay sober. As suggested above, each recognizes the other's weaknesses, and each plays a part in the total support system that has developed around the Thornton Freewill Baptist Church and "The Words of Love Broadcast."

In 1984, at Silas Creek Union Baptist Church in Lansing, North Carolina, I heard an aging mountain preacher talk to his church about congregational unity, employing in the process the teepee metaphor: Each pole stands because it leans against others, the entire collection giving the conical structure strength.[34] My impression is that Brother Dean Fields would affirm the message of that metaphor, recognizing at the same time that he and his congregation personify both the strengths and weaknesses of the principle: These highly vulnerable people support each other, but they may not be particularly strong apart from the unit. My cursory exposure to the group has provided little that would allow me discuss that last question, except perhaps to say that like most humans they undoubtedly struggle with their own private sets of "ghosts" and "devils."

I will close this chapter on one very positive note, observing that as far as I could tell these people have been good to each other—no noticeable church squabbles, demonstrations of jealousies, or hostile competitions. The open, warm, and generous nature of Brother Dean Fields appears to have established a tone with which everyone else harmonizes. Furthermore, I do not have any trouble accepting Fields's judgment that the Thornton Creek community is better for the Thornton Freewill Baptist Church's being there. Obviously, he also feels that Letcher County, Kentucky, is better for "The Words of Love Broadcast" being aired.

Sister Brenda Blankenship, WELC, and Women in the Airwaves of Zion

> And it shall come to pass in the last days, saith God, I will pour out of my Spirit upon all flesh; and your sons and your daughters shall prophesy, and your young men shall see visions, and your old men shall dream dreams;
>
> And on my servants and on my handmaidens I will pour out in those days of my Spirit, and they shall prophesy.
>
> Acts 2:17–18

Welch is the seat of McDowell County, West Virginia, one of the most economically depressed coal-mining regions of Appalachia. The town lies at the confluence of Elkhorn Creek and Tug Fork, tributaries of the Big Sandy River, a northwest-flowing waterway that drains watersheds from Virginia, West Virginia, and Kentucky into the Ohio River; and the municipality has become the home of WELC-AM, an airwaves-of-Zion station that, early in the fieldwork for this study, attracted my attention because of its heavy distribution of female exhorters. In this chapter I examine one of these preachers, Sister Brenda Blankenship, a "full-gospel" Pentecostal evangelist from Premier, West Virginia. I also focus on aspects of the McDowell County/WELC situation that suggest a religious environment heavily influenced by women.

The Community and the County

By the mid-1800s this area of Elkhorn Valley, West Virginia, had been sparsely settled and marginally cleared, but the town was not really developed until after 1888 and was not incorporated until 1894.[1] All of this occurred after the arrival of a railroad that changed the region's economic base from subsistence farming to commercial logging and coal mining. At the time of its incorporation, the community was named after Captain Isiah A. Welch, a former Confederate officer responsible for much of the initial development of the area.[2]

Prior to 1888 all of McDowell County was sparsely populated, but that year the Norfolk and Western Railroad, approaching from Bluefield to the east, completed a 3,100-foot tunnel through Flat Top Mountain and brought tracks to Elkhorn Valley. During the next four years that line was extended northwestward through McDowell, Mingo, and Wayne counties, eventually reaching Kenova, West Virginia, on the Ohio River. This rail transportation corridor gave the regions of southwest West Virginia access to surging industrial markets of the upper Midwest, spurred logging operations, and engendered rapid development of the Pocahontas-Flat Top coalfields.[3]

According to Ronald Eller, a historian of industrialization, "Between 1880 and 1900, the population of the county [McDowell] increased by over 600 percent, and coal production reached over 4 million tons per year."[4] McDowell quickly became the most productive coal county in the state,[5] and Welch emerged as the county's trading center, drawing to its compact municipal center a wide array of business and financial establishments.

During the late 1930s a corps of researchers working for the Writers' Project of the Works Progress Administration visited Welch as part of their statewide travels requisite to producing *West Virginia: A Guide to the Mountain State*. In that work these writers recorded the following about the booming town:

> The business section occupies a level area in the narrow Elkhorn Valley, and the residential section is scattered on surrounding hills. Welch, the trade center for the surrounding coal fields, is served by the Norfolk and Western Railway. Narrow streets provide a traffic problem, and except for the pleasantly suburban residential area, Welch is a congested town, so much so that it has been called "Little New York" by the *New York Times*.
>
> On Saturday nights thousands of coal miners from dozens of operations descend upon the town to shop and find amusement.[6]

As the town grew into a city and as money flowed through its numerous commercial establishments, Welch attracted a mercantile-managerial-professional class that in the 1940s and 1950s provided the community a more "cultured" image than that held by other towns in the county. Functioning both as a trade and service center, Welch developed all the institutions necessary for a modern urban society. Thus by 1959 the local chapter of the Daughters of the American Revolution could brag about the city's "three large hospitals, . . . the only daily newspaper in the county, . . . ten churches, fine schools, three service clubs, women's clubs, war service organizations, a public library, and two chambers of commerce."[7]

The rapid population growth that Welch experienced after 1888 continued during the first half of the twentieth century, peaking during World War II, when mines and processing plants operated around the clock so shift workers could produce vast tonnages of coal for the war effort. Throughout the early years of this boom, communities in the Elkhorn Creek and Tug Fork valleys were packed with a citizenry that was young and heavily male, as a labor-intensive industry brought thousands of mine workers to the region—a good portion of them being foreign immigrants—crowding the boarding houses to overflowing with either unmarried miners or miners who had left their families outside the valley, or—as was the case for Eleanor Parker's father—outside the country. From those boom-time war years until the present, however, Welch has suffered consistent losses in its business and industrial base and has experienced radical changes in the nature of its population.

From the 1950s through the 1970s, numerous large shaft mines and coal-processing plants along the Elkhorn Creek and Tug River valleys closed, leaving a sizable portion of the active mining operations of the 1980s in the hands of "strippers," who use an abundance of machinery but few workers. Thus thousands of the miners who once jammed the town's streets on Saturday nights sought employment elsewhere during the 1960s, 1970s, and 1980s—or fell onto pension or welfare rolls. The 1990 United States Census figures show the municipality to have a population of only 3,028, down 54.2 percent from the town's highest census total of 6,603 (1950).[8]

Perhaps more telling is that the people of the surrounding area have grown poorer, older, and more heavily female. In 1961, for example, McDowell County became one of only eight regions of the nation initially to be served by President John Kennedy's newly instituted food stamp program. According to a public relations document distributed by the town of Welch, "Six hundred county families appeared during the first three days of registration."[9] In addition, census figures have shown a steady rise in the average age of county residents and a gender distribution that is becoming more unequally female. These changing demographics appear to result, in part, from so many of the people that remain in the region being aging female pensioners or younger female heads of single-parent households. In 1940 only 1.9 percent of McDowell County's population was age 65 or older; whereas, by 1980 that population segment had grown to 10.3 percent. Furthermore, in 1940, 47 percent of McDowell County's population was female, but that number had grown to 51.5 percent by 1980. Finally, the county's percentage of families below the poverty line in 1980 was 19.3, compared with the national average that year of 9.6 percent.[10]

During the last forty years McDowell County has fared even worse than Welch in terms of population decline, dropping from a 1950 high of 98,887 to only 35,233 in 1990, a total decrease of 64.4 percent.[11] This downward spiral in population is attributable both to a regression in the region's coal production and to the changing methods of operation in that industry. McDowell County's yearly coal tonnages peaked during World War II and then began to drop during the late 1940s (from 27,588,505 tons in 1941 to 20,912,416 tons in 1955[12]). Furthermore, surface mining—low in labor usage as mentioned above—increased substantially during those postwar years (from 8.8 percent of West Virginia's total coal output in 1950 to 20.7 percent of the state's coal production in 1980[13]). "Surface mining and the mechanization of underground mines," says historian Otis K. Rice, "drastically reduced the labor force in the West Virginia coal industry."[14] "In 1950," notes another source, "a tipple work force was made up of 12 men. In 1970 it used only two."[15] "From 125,699 in 1948," adds Rice, "the number of miners dropped to 41,941 in 1969."[16]

One result of this 66.7 percent decline in coal industry jobs in just twenty-one years was that McDowell County communities such as Welch, Keystone, Gary, War, English, Iaeger, and others were plunged into an economic environment far less healthy than the one that earlier had existed, producing not only abandoned mining sites like those described in the last chapter but also hundreds of closed and boarded-up commercial establishments. All of this in turn precipitated the flight of the working-age population, particularly males. Welch, West Virginia, is much like Neon, Kentucky, only on a larger scale: employment opportunities are few, and each year the residents appear to be growing older, poorer, and more heavily female.

Here and there in Welch are evidences of individual and group efforts to preserve, restore, or rebuild, especially among buildings and other structures under the control of the municipal government, and also in some of the elite residential areas—the old homes that hug the higher street levels on the mountainsides (once residences of mine

McDowell Street, Welch, West Virginia, 1946.
Courtesy of The National Archives.

McDowell Street, midmorning of a weekday in August 1991.

managers, professionals, or well-to-do merchants). At the time of this writing, Mayor Martha Moore (elected in 1986) is struggling to preserve for her town an image of life and vitality, and she fiercely promotes the city and region as viable and economically rechargeable.

During her administration, Moore has arranged to have several abandoned and decaying structures that stood within the city demolished and removed from the landscape. In addition, she has made certain that the streets of Welch are clean and inviting, even when they run in front of blocks of storefronts largely devoid of commerce. For example, in 1987 the Howard Street section of Welch was refurbished, producing an attractive walkway, park benches, trees and shrubs, and even a reviewing stand, all efforts at bringing some civic pride back to the town. In another action, Moore instituted the Welch Advisory Corps,

> formed and modeled after the "New Deal" of Franklin D. Roosevelt. It advocated a three-way plan to solve city problems. . . . The Public Works Administration (P.W.A.) would coordinate volunteer efforts to revitalize the city. The Welch Civic Conservation Corps (C.C.C.) to involve area youth and church groups for city beautification. The National Industry Recovery Administration (N.I.R.A.) to interest out of county manufacturers to choose Welch for plant locations.[17]

Moore wants the world to know Welch is not dead, just wounded. Furthermore she wants Welch kept in perspective with other American cities, and the citizens of Welch to be recognized as what she says they really are—fighters. "We're not the only urban area having trouble," she declares. "Look at the Northeast and the industrialized Midwest. The biggest problem we have to overcome is the stigma that has been laid upon us, that all this is the result of some deficiency in our character. That's just not the case."[18]

Were I a resident of Welch, West Virginia, I would lend my energy to this woman's causes, challenging though they are. On the morning of August 1, 1991, for example, I caught a glimpse of one feature of Welch's problem, the heavy dependence upon welfare benefits and pension payments. It was about 8:30 in the morning and I was in the West Virginia Cafe on McDowell Street, that once heavily trafficked main thoroughfare of this city—the only customer in the restaurant.

When I paid my check, I spoke to the owner of the cafe, Margaret Henrichs: "Is this your typical morning business?" I asked.

"No, today's not a usual morning," she answered. "Today's the first of the month. They're all down at the post office right now, picking up

Welch, West Virginia, August 1991.

checks. Then they'll go to the banks. By ten o'clock there'll be folks on the street. Then business will be good for about five days. Tomorrow will be booming."[19]

Margaret Henrichs was right: By 10:00 A.M. there were people on McDowell Street. Not many, however, and they included what seemed to be a heavy concentration of elderly people. Nevertheless, several young women with small children did appear, most of them entering the street's Dollar Store, apparently to purchase school clothes. This moderate surge of business still produced only a trickle of street traffic, not at all comparable to the congested flow of people and vehicles shown in the 1946 photograph in this chapter. By 10:30 on that Thursday morning so sparse were both the pedestrian and automobile traffic on McDowell Street that I felt the day must have been Sunday or that the time of day was much earlier than the hour indicated by my watch.

Elkhorn Valley

A traveler approaching Welch from the east on Highway 52 will cross Flat Top Mountain and drop down into Elkhorn Valley, moving in succession through the communities of Maybeury, Elkhorn, Northfork, Keystone, Vivian, Kimball, and Superior, all old mining-camp locations, and all in varying stages of disrepair. Throughout the valley there are complex structures—commercial, residential, or industrial—many built of brick or stone in the 1920s, 1930s, and 1940s—that stand abandoned—hopelessly irreparable, precariously unstable, and depressingly unsightly. Too often the images are of decay and near desolation, perhaps viewable as some of the worst examples of the aftermath of Appalachian industrialization.

My first visit to this valley was in February 1989, and the winter barrenness of the mountains lent a particular starkness to the environment: absent the lush greenness of an Appalachian summer, the mining-debris-strewn landscape looked doubly desolate. From late spring through fall a heavy canopy of either dark green or autumn-colored foliage conceals much of the smaller upper-hillside derelictions; but the larger abandoned buildings on the lower ascents or on the valley floor are not concealable, lending, year-round, at least a partial ghost-town image to all the communities between Maybeury and Welch, as well as elsewhere in the county. Few regions in Appalachia communicate a sense of commercial depression, industrial ruin, and environmental exploitation more palpable than that found in numerous areas of McDowell County.

Even Elkhorn Creek becomes a strong symbol of this state of desolation, occasionally littered not only with industrial discards but with years of domestic waste—perhaps an old home appliance, a rusting section of a bed frame, an automobile tire, the wheel-less body of an old tricycle, or a host of other objects that seem to suggest a departing working class of people saying, "The companies that employed us left their wreckage behind: why can't we?"

Within the city of Welch, action has been taken to clear debris from the beds of Elkhorn Creek and the Tug River, one of the projects of Mary Sidote, wife of the owner/manager of WELC.[20] However, that action has not always been emulated in other areas of the county, and waterways are polluted not only by old mine runoffs but by all those man-made objects people find no longer usable, objects that in some instances have been deposited in these waterways by one of the several floods that have visited the valley.

Suffice it to say that much of the old McDowell County must be cleared away before any new McDowell County can emerge. The challenge is enormous and will test the indefatigable spirit, endurance, and sociopolitical ingenuity of the Martha Moores of the region.

WELC

On the air since August 19, 1950, station WELC broadcasts from studios that lie on the west side of Welch, off Highway 52 as that road winds toward Iaeger in McDowell County and Williamson in Mingo County. The broadcast facilities are housed in a three-story structure cradled within a cluster of trees that cling to the steep incline of Premier Mountain. Much higher, on the ridge of this mountain, rests the WELC transmission tower, a structure that each Sunday sends forth along the airwaves of Zion an extensive program of locally produced live religious broadcasts heavily imbued with the voices of West Virginia women.

These live religious programs originate in an add-on studio attached to the second story of the west end of the building, an approximately sixteen-by-sixteen-foot space that houses an upright piano, a speaker's stand that usually faces this piano, two microphones (one at the piano and one at the podium), a number of metal folding chairs (frequently occupied by visiting broadcast observers), a large clock (synchronized with the main studio clock), an "On The Air" sign, a Welch Insurance Agency wall calendar, and a wooden cross, overlaid with sequins, that hangs behind the piano, becoming the identifying icon for this space. The cross, I was told, was a gift of one of the Sunday airwaves-of-Zion groups, and this studio is viewed as an airwaves-of-Zion sanctuary.

WELC was recently awarded a spot on the FM band, and this building accommodates separate AM and FM control rooms that adjoin each other. The closeness of the two consoles allows WELC owner/manager Sam Sidote—who during the Sunday-morning programming operates both boards—to dart back and forth between the two panels of switches and dials, saving himself some personnel costs in the process. On two Sunday mornings I interviewed this man while he was engaged in all these back-and-forth maneuvers, and on both occasions I was impressed not only with his energy but with his degree of concentration: including my questions, he had three things going at once.

During most of its broadcasting week, WELC simulcasts the same programming on both its frequencies. On Sundays, however, and again for brief periods on weekday mornings, there is religious programming that goes out over only the AM band, leaving the FM band for those residents of McDowell County who choose not to tune to the airwaves of Zion.

"We've got a lot of faithful listeners who want those broadcasts preserved," says Sam Sidote. "The AM frequency allows us to do that."[21]

As of this writing, Sam Sidote is seventy years old, a small, wiry man—gray-haired, but explosively energetic and seemly inexhaustible, loaded with enthusiasm for the town of Welch, WELC, the people who work for him at the station, all of the groups that come in for Sunday programs, and such West Virginia and national politicians as agree with his ideologies. Operating as a committed publicist for Welch and the region, Sidote argues that McDowell County can be reborn if people will just "get off welfare and get back to work."[22] It's only a matter, he believes, of persistent industry and faith, a power-of-positive-thinking philosophy he consistently espouses.

A genuinely warm and likable man, Sam Sidote exudes an indomitable spirit that matches the positive and progressive attitudes of Mayor Martha Moore, but that occasionally seems to belie the economic and social realities of Elkhorn Valley: he has built a profitable radio station, so why cannot others rejuvenate their respective parts of the city or county? He sees no intrinsic obstacle to such a goal. Like

WELC, Premier Mountain, Welch, West Virginia.

Sam Sidote at the WELC controls.

Mayor Moore, he believes Welch, West Virginia, can be reborn if the residents just stay with the task. Hard work and a generous spirit, he says, will prevail over all odds.

It is obvious that Sam Sidote takes pride in Radio WELC. The exterior walls of the multilevel concrete-block structure are painted white, and each window is equipped with a green and white metal awning, all looking fresh, clean, and cared-for. Centered on the exterior face of the structure's third floor are the letters "RADIO WELC," and down by the highway there is a sign that again displays the station's call letters. From this spot a neatly trimmed low hedge leads up the driveway and curves around to the east end of the building where a downlink dish is positioned to receive syndicated programs via satellite transmissions. In many respects WELC appears atypical to the main body of commercial establishments in Welch and the remainder of the county.

The Sisterhood of WELC's Airwaves-of-Zion Programming

When I first visited WELC, on February 12, 1989, the Sunday programming included eight airwaves-of-Zion broadcasts, starting at 9:00 A.M. and continuing—with only a few breaks in the schedule— until 3:00 P.M., providing in the process one of the heaviest concentrations of women radio preachers I experienced during this study. At 9:00 Sister Ann Profitt of North Spring, West Virginia, sang and preached for thirty minutes in her "Jesus Is the Light of the World Program," accompanying her hymns with a guitar. She was followed at 10:30 by Sister Brenda Hall of Davy, West Virginia, with her "Heart to Heart" broadcast. After a fifteen-minute break in the programming, Sister Charlotte Bolden of Twin Branch, West Virginia, opened her "Pentecostal Full Gospel Broadcast" at 12:15. Finally, at 12:30 there was "The Evangelist Brenda Blankenship Broadcast," the program featured in this chapter.

Forty-four years of age at the time of this writing, Blankenship, a Pentecostal evangelist from Premier, West Virginia, produces her program with the help of her husband, Jimmy Blankenship; her sister, Dora Justice; another female supporter, Sandy Riffe; and Tom Christian, a musician and singer who occasionally helps out with other WELC airwaves-of-Zion broadcasts.

Of these four helpers, Justice is probably second to Brenda Blankenship in program involvement. She is a singer and testifier/ exhorter, this combination term signifying her tendency to move from

165

a "this is what God's done for me" testimonial into a passionate rhetoric that becomes preaching. Justice's main contribution to the broadcast, however, is as a singer, either solo or in harmony with Blankenship.

Sister Sandy Riffe is a more subdued participant than either Blankenship or Justice, remaining largely in the background when not singing, usually some distance from the microphone, but—in response to the contributions of others—waving her arms, swaying, softly clapping her hands, and testifying or praising quietly to herself. Through these actions she aids both Sister Blankenship and Sister Justice, serving them as a highly empathic respondent.

Somewhere in his mid- to late-fifties, Brother Tom Christian plays his guitar not only for Blankenship's radio broadcast, but also for her tent revivals, her home services, and her street preachings. He is an exuberant and very audible "praiser," throwing out those short exclamations that are so typical to this genre of religious expression: "Praise God," "Amen," "Bless her, Lord," "Glory to Jesus," and the like. His contributions add considerably to the overall intensity of the broadcast, and Brenda Blankenship views him as an indispensable part of her team.

"I don't know what I'd do without Brother Tom," she declares. "He's a true man of God. He knows what Spirit is, and he gives his all."[23]

Christian fits well the pattern of one type of performer often seen in airwaves-of-Zion settings, the strong supporting actor who not only contributes but activates others, sparking and intensifying the main performer or performers. It seems doubtful that Brother Tom Christian would ever establish his own program, but he works well in tandem with a leader.

Jimmy Blankenship is usually the most subdued of the five broadcast participants, a relatively quiet man whose natural inclinations are to play only supporting roles. A former heavy-equipment operator who suffers some kind of medical disability, Brother Jimmy Blankenship only recently joined his wife's evangelistic efforts, having been "saved" the year before my first visit to WELC. As a result of this newness to the faith, he still finds it difficult to be as demonstrative in the expression of his religious feelings as are the three women and Brother Tom Christian. His service to the broadcast is as a supporting singer and occasionally as an announcer. Sister Blankenship, however, always leads the program, controlling the thirty minutes of evangelism with a zeal, determination, and energy that are difficult to surpass or to match.

Brother Tom Christian at WELC.

Although there are male participants scattered throughout WELC's Sunday morning and afternoon programming (preachers Dewey Russ, Curtis Cantrell, Ronnie B. McKenzie, and Gene Ball, as well as several male supporting performers), the female airwaves-of-Zion participants at WELC take focus, in terms of authority, numbers, and level of exuberance. So strong is the female involvement that this segment of WELC's airwaves-of-Zion personnel assumes the aura of a sisterhood.

Three of the women preachers (Hall, Bolden, and Blankenship), are assisted by other females who exhort, testify, sing, pray, or play musical instruments, and in general set the exuberant energy levels of this WELC programming. During my first visit to WELC, Sister Brenda Hall's thirty minutes of broadcasting featured the preaching of Sister Ruby Chaffin, a woman in her sixties who served as the assistant pastor of the Marytown Holiness-Pentecostal Church. Sister Charlotte Bolden was helped in her program by the piano playing and singing of Sisters Barbara Bolden and Josey Greene, both from the Big Jenny Pentecostal Full Gospel Church. Then, as noted above, Evangelist Brenda Blankenship was assisted by Sisters Riffe and Justice. Other women figured prominently as members of the studio audience, emotionally responsive through arm waving, shouting, semi-audible praying, and background singing. This was particularly true during the broadcasts of two of the male preachers, Gene Ball and Curtis Cantrell.

Finally, there is Sister Ann Profitt's solo performance in her "Jesus Is the Light of the World Program." Sister Profitt devoted several years to assisting her husband as he directed this particular thirty-minute broadcast, but when he died—some time before my first visit to WELC—Sister Ann dutifully stepped in and kept the program going, filling her half-hour with singing, announcements and recognitions, prayer, and preaching. The three times I visited her show she was wearing the same red polyester jersey preaching gown, decorated with a gold-colored brooch that spells out the name "Jesus."

There is much about Sister Ann Profitt that makes me think of her as a female counterpart to Brother Kelly, described in chapter 4. She's an aging solo exhorter, dedicated to a tradition that undoubtedly has great meaning for her, and persisting in the continuation of that tradition probably with some degree of personal sacrifice.

Exemplifying the freedom of religious expression and leadership that females possess in the various Holiness-Pentecostal fellowships of Appalachia, women of the WELC airwaves-of-Zion family practice

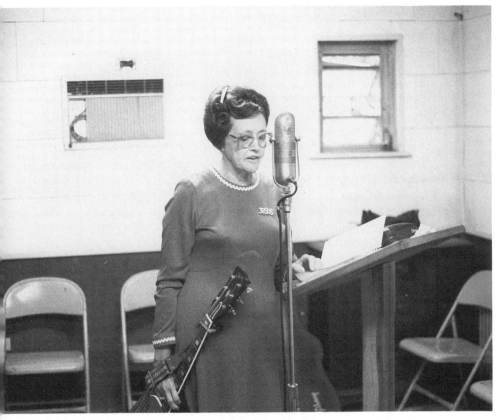

Sister Ann Profitt at WELC.

their callings with confidence, assertiveness, and fervor, usually providing the main drive and command necessary to the production of any one program. The husbands of Brenda Hall and Brenda Blankenship, for example, operate only in obvious support positions, assisting in bringing to fruition the evangelistic visions of their respective spouses. Furthermore, it is difficult to ignore the larger roles these women play in their respective religious communities: Brenda Hall serves as pastor of Marytown Pentecostal Church; Brenda Blankenship promotes herself as an evangelist, accepting calls to conduct revivals not only in West Virginia, but also in Kentucky, North Carolina, and Tennessee;[24] Ann Profitt apparently shoulders responsibility for all aspects of her deceased husband's ministry; and both Charlotte Bolden and Ruby Chaffin assume pastoral duties in McDowell County churches. This is that sisterhood of WELC's airwaves of Zion, and on Sunday mornings and afternoons these women fill this station's small production studio with their urgent calls to redemption and piety, perhaps demonstrating to their respective audiences that they (the sister preachers) fulfill the promise of Joel's prophecy, quoted in Acts 2:17–18:

> And it shall come to pass in the last days, saith God, I will pour out of my Spirit upon all flesh: and your sons and your daughters shall prophesy, and your young men shall see visions, and your old men shall dream dreams.
>
> And on my servants and on my handmaidens I will pour out in those days of my Spirit; and they shall prophesy.

The "last days" phrase of this Scripture signifies for Brenda Blankenship and her fellow preaching sisters the second Pentecost (the "latter rain"), which they believe either to be in progress or imminent. Furthermore, the sisterhood defends this Scripture's sanctions over the prohibition said to be embodied in 1 Corinthians 14:34:

> Let your women keep silence in the churches: for it is not permitted unto them to speak; but they are commanded to be under obedience, as also saith the law.

"If the Lord has told me to preach," says Sister Kathy Benfield of Drexel, North Carolina, "I've got to preach, whether everybody thinks it's proper or not."[25] Brenda Blankenship echoes this sentiment, but adds that the men just do not understand the situations to which Paul's directive applied:

The women kept talking out loud in church, leaning over and asking their husbands something about the lesson being taught, and Paul got tired of that and said that the women should keep quiet in church and if they wanted to know something they should ask their husbands at home. That's just being considerate of others. He didn't say the women shouldn't preach.[26]

One goal of this chapter is to focus a more concentrated attention upon these women preachers of the airwaves of Zion. The Sunday programming at WELC encourages such a focus, and the growing percentage of women within the McDowell County population may suggest the interaction of certain sociocultural variables. This is a question to which I will return later in this chapter when I examine a tent revival with which Sister Brenda Blankenship was involved.

The Radio Preaching of Sister Brenda Blankenship

My previous studies of highly traditional Baptist subdenominations have not provided the opportunity to examine the preaching style of a woman, so determined these old-time Baptist groups have been to preserve the male purity of their ecclesiastical ranks. Holiness-Pentecostal environments, of course, change that situation, exposing the interested religious ethnographer to the hortatory talents of such women as Sister Kathy Benfield of Drexel, North Carolina, and Sister Brenda Blankenship of Premier, West Virginia.

Brenda Blankenship has been preaching for twenty-five years, "ever since when I was saved at nineteen," she says. Her enthusiasm, at times, has been threatened; but she claims to have maintained a consistency in her evangelistic drive. "I've been through some tough times," she observes,

but I always kept up my work for the Lord.

I lost my nineteen year old daughter in 1983, because of a drunken driver. It liked to kill me. That took something out of me for a while, but I come back stronger. Praise God! It helped when Jimmy got saved, too. Praise the Lord!

I pastored a church from 1979 to '83. Then my dad died in '82, followed by my daughter the next year. Next one of my sisters died of cancer, causing me a lot of sorrow. We was real close.

All that made me want to go back to my evangelizing. Evangelizing brings real meaning to my life. I feel more alive in that work than in pastoring.

I want to get me a tent—maybe a small one, that'll fit maybe two hundred. That 'bout all of a tent I could haul about. And I'd need some help at that. I can't handle no big rig like some preachers have.

I want to keep up my street preaching, too. Dora and me—and sometimes Tom Christian—go down on a street, maybe in Bradshaw or Roderfield, and just start preaching and singing. Dora carries her little electric organ. People are used to this now, and we get some nice crowds.

Then I have lots of home services. People just invite me over, and we sing, and pray, and preach. There'll be several families there, mostly women and children, but some men.

I'm gonna get me a cordless mike for my preaching. They're handy. Lets you to move about the way the Spirit directs.

I especially enjoy my radio preaching. That allows me to reach lots of people. It's not for me—the glory. My main goal is preaching Jesus.[27]

Like Sister Benfield, Blankenship holds nothing back when she preaches. Her delivery is forceful, rapid, emotional, explosively varied, and exhaustively physical. Throughout her exhortations she shouts, whoops, hollers, drops into glossolalia, throws her arms and head back and up, quickly stamps her feet, jumps, sprints, and joyfully exclaims time after time, "Praise God," "Amen," "Glory to God," or "Hallelujah!"

Her dynamism appears to exceed those levels exhibited by Brothers Johnny Ward, Dewey Ward, and Dean Fields, if for no other reason than that she is small, lithe, and dartingly quick, ready to spring with impressive suddenness into any posture required to communicate her soaring and seemingly ungovernable and limitless spiritual passions. In summary, she is as forceful and energetic a preacher as I have ever witnessed, constantly in motion and constantly employing her deep and somewhat husky voice to roam the available ranges of emotional expression. She may lack some of the rhetorical fluency possessed by more educated exhorters, but she certainly is not lacking in vivacity.

When I first witnessed the radio preaching of Sister Brenda Blankenship, I thought the small studio at WELC could not hold her, that she might grab the microphone and rush out into the corridor that leads to the control room where Sam Sidote sits. Within the limits of the room and her mike cord, she surged from side to side and forward and back, occasionally performing a skipping step while her body was held low and crouching, at other times simply holding in one spot while bouncing on her toes.

"I really like to preach outdoors or under a tent," she told me again.

"You've got more freedom, more space to move about. That suits the way I preach, and that's why I want that cordless mike."[28]

Throughout such moments of physical exuberance, Blankenship pours out her words in furious bursts of rhetoric, broken occasionally by one of her whoops, hollers, or shouts. Her whoops sound like the word itself sounds—sharp, sudden, and jubilant, an expression of immense exhilaration; her hollers are elongated and more wailful in tone, a rising inflection that wavers as it soars; and her shouts are somewhat louder and considerably more explosive, forceful, though not angry.

As is traditional for much of Appalachian preaching, she begins her exhortations in prose but drops into a rhythmical chant once she reaches the quickened pace that she says flows from "anointment," the condition of spiritual empowerment that comes as a result of the "possession by the Holy Spirit." Under "anointment" she not only speeds her delivery, but she embellishes that delivery with a lyrical style that rises to poetic expression. Rhythm, of course, is the main factor in this process as her Appalachian homiletic chant develops.

Blankenship's style mirrors well that traditional Appalachian mode of sermonic delivery, except in one respect. At the end of each linear segment she does exhale and quickly inhale, but not with the explosive "haah" that characterizes much of the region's old-time Baptist pulpit exhortation. The rhythm is there, but without the strongly accented punch at the end of each line. Instead she frequently employs the closing or transitional "amens," "praise God's," and "hallelujahs," so common to Holiness-Pentecostal preaching.

Once Blankenship really gets her preaching under way and has dropped into her rhythmical pattern, she seldom allows any space to develop within the flow of words and other sounds. A continuous outpouring of impassioned expression assaults the microphone and seems almost to pulsate within the confines of WELC's not-so-acoustically-controlled studio. In addition, Brother Tom Christian keeps up a steady flood of supportive rhetoric, a repetition of traditional exclamations, occasionally serving as counterpoints to Blankenship's homiletic measures.

Blankenship's moments of glossolalia are interesting, in the sense that like her "amens" they also appear to serve as transitional passages. Seldom will she break a developing rhythmical segment by dropping into "tongues," in part—it seems—because the glossolalia changes the cadence of her delivery, moving the flow of vocalizations into a format that sounds more like prose. Instead, she waits until she reaches the close of a peaking sequence of chant and then comes out of that sequence with a quick burst of "tongues."

The following is a transcription of the opening moments of Blankenship's preaching on February 12, 1989. In the final moments of this sermon segment she established an unbroken rhythmical pace that gradually climbed in intensity:

> I'd like to take my text today from the twenty-fifth verse [of Hebrews 7], says "wherefore He is able to save" us "to the uttermost [*dramatically rising in pitch*] that come unto God by Him." Amen! . . . My text will be today "wherefore He is able." [*She claps her hands and intensifies her delivery in doing so, dropping in the process into the beginnings of her rhythmical chant.*]
> Sister Sandy sing that song, Amen!
> "He's a great big God, Jehovah,
> So greatly to be praised,
> A God [*suddenly accented on* "God," *with a rapid soaring in inflection*] that'll carry us over
> Every mountain sin has raised.
> Out of all the gods men's worshiped,
> there's a difference we can tell.
> He's a great big . . . [*she emits a loud* "Whoop!" *and a* "Hallelujah," *breaking the flow of the verse and seeming to forget where she is in the lyrics of the hymn*] . . . born God's only son,
> a God [*elongated and wavered*] that cannot fail."
> Amen! Man may fail. Amen!
> But God never fails.
> The Bible says here in the twenty-fourth verse
> that He has an "unchangeable priesthood."
> And we see people, maybe today,
> that'll do wrong things
> and cause them to change to something else.
> But Jesus never changes.
> He's the same yesterday, today, and forever!
> And I thank God today, Sister Dora, amen,
> that God . . . [*suddenly breaking the thought*]
> I thought about this Scripture after we talked last night.
> Amen!
> "wherefore He is able to save them to the uttermost that come to God by Him." ["Him" *is emphasized, again by an elongated wave as Blankenship bounces on her toes.*]
> Today man is seeking man's help, and that's good to have one of those prayers. [*Again she claps her hands in a rapid fashion, whirling halfway around as she does so before quickly whirling back.*]

Because I thought, Brother Tom,
the Lord teaches us to pray
one for another.
But, oh, the Lord said
that He is able to save.
We can go to one another, amen,
but I can't save you,
and I can't deliver you
out of your troubles. [*At this point she emits an elongated* "Whoop!"
   and again bounces up and down.]
But David said in Psalms 34 and 4, ["Whoop!"]
"I sought the Lord."
I sought Him! ["Whoop!"]
I didn't seek my neighbor;
I didn't seek my friend.
I sought the Lord!
And he heard me.
And delivered me ["Whoop," *with accompanying bounces.*]
from all my fears.
Glory be to God! Hallelujah!
It said,
"wherefore He is able . . . [*a soaring emphasis on* "able"]
to save them
to the uttermost"!
Uttermost means ["Whoop!"] hallelujah,
you can't get around Him [*beginning to dart from side to side*]
and you can't get over Him,
and you can't get under Him!
Uttermost means to the very top.
Amen, there is no top.
You'll see that river.
He said, "river that could not be passed over."
Amen, to the uttermost! [*A shouted phrase, ended by a* "Whoop!" *and*
   more jumps.]
Amen, hallelujah!
Amen, He's able to save those
that come unto God by Him,
"seeing He ever liveth
to make intercession for them."
When man's sleeping,
and they're not thinking about praying
for those prayer requests you put forth,
the Bible said Jesus ever lives [*a soaring inflection on* "lives," *her*
   upper torso arched sharply backward]

175

"to make intercession." [*Phasing out with a quick burst of glossolalia,
lasting no more than about three seconds, she now was ready to
begin another acceleration in chanted rhetoric.*][29]

One of Blankenship's techniques is to build intensity through the
use of repetition, establishing a theme line that alternates with a varied line. During these moments she may shut her eyes, hold the mike
close to her mouth with her left hand, raise her right hand high above
her head, toss her head back, and—in rhythm with her own words—
repeatedly bounce on her toes, waving that right hand as she does so.

> Like for the children of Israel,
>   He'll bring you out of bondage.
> Your home may be a wreck;
>   He'll bring you out of bondage.
> Your children may be going wrong;
>   He'll bring you out of bondage.
> Your husband may lose his job;
>   He'll bring you out of bondage.
> Your health may not be right;
>   He'll bring you out of bondage.
> You may be drinking;
>   He'll bring you out of bondage.
> You may be lost in sin;
>   He'll bring you out of bondage.[30]

A second technique she employs quite frequently is to engage in
brief mock dialogues with a cynical "sinner," in the process joyously
laughing at her own cleverness. During a revival sermon she preached
in a Pentecostal church in Kernersville, North Carolina, Blankenship
began speaking about why many people are in hospitals. The illness
she wanted to address was "nervousness," a mental state she felt was
often brought on by an "unsaved" life. "A lot of people that are in the
hospitals today are there for one reason," she said, "their nerves. I've
walked into hospitals and people will be half out of their minds, with
nerves." Then in closing this sermon segment she threw out the following question, as if asked derisively by an unbeliever: "Are you the doctor?"

No, but I come from the Doctor. [*Smiling triumphantly.*]
His name's Jesus. [*Shouted joyfully.*]
You don't need no prescriptions here. [*A "whoop!" and laughter from her
audience of believers.*][31]

Blankenship enjoys telling sermonic stories, some apparently rooted in fact and some apparently wholly fictitious. These narratives often involve struggles with demonic power, since she believes the devil to exist as a natural enemy of God and Jesus. Occasionally, either within the story or at its close, she engages in imagined dialogue with Satan, building pride, forceful determination, and merriment in her audience as she mocks and outwits the evil trickery in which Satan has been involved. At the close of such sequences she invariably pulls the audience into her narrative, using "we" and "us" to make her story the audience's story, and her rhetorical victory the audience's victory. In a climactic declaration, that often brings applause and shouts when she has a congregation present, she typically exits such scenarios with a line similar to the following: "He tried to impress us, depress us, and oppress us; but we rebuked him. Hallelujah, we rebuked him! Let me hear an Amen from those who rebuke the devil."

In notes I recorded after my first exposure to the preaching of Sister Brenda Blankenship, I wrote the adjectives "paroxysmal" and "convulsive," seeking to remind myself of the sudden, explosive bursts that frequently occur in this exhorter's performances—explosions in both vocalics and kinesics, explosions that might send her charging across an open space or that might find expression in her whoops and hollers and exclamatory "hallelujahs," explosions that galvanize the emotions of her listeners, often sending them into their own peaks of joyous verbalization. "Untrammeled" and "boundless" are also appropriate adjectives with which to characterize the limitless range of emotions this woman employs to communicate her affective states, her convictions, and her evangelistic urgings. Once again I will simply observe that she "holds nothing back"—no reserves, no moderation.

Hymn Singing during the Broadcast

Singing constitutes approximately two-thirds of Sister Blankenship's broadcasts. At the beginning of the program a prayer is offered, and a few announcements may be made, but Blankenship does not include in her program format any wide range of people recognitions like those—discussed in chapter 1—that became a part of Brother Hall's WAEY broadcast. Instead Sisters Blankenship, Justice, and Riffe—assisted by Brother Christian—devote approximately twenty minutes of this half-hour broadcast to an exuberant style of singing that becomes an emotional build-up to Blankenship's preaching.

As mentioned earlier, there is a piano in the WELC studio, and both Sister Brenda and Sister Dora play this instrument with considerable energy and flair, emphasizing volume and occasionally covering a transition in the music by executing a dramatic running of the keys. Furthermore, Blankenship and Justice constantly team together to sing for tent revivals and other such services, and they take great relish in performing loud, exuberant, and fast-paced gospel songs that have that decidedly Pentecostal sound.

> Send down the rain, Lord!
> Send down the rain.
> Send down the Holy Ghost,
> Holy Ghost rain.
> Send down the rain, Lord!
> Send down the rain.
> Send down the Latter Rain.

It is not unusual for these numbers to become explosively joyful, with one of the women beginning to shout as the other continues the hymn, with all members of the groups clapping in rhythm with the song. At such times Brother Tom Christian also comes in with his strings of "praising" exclamations, building the energy level to an even higher plane. Then the singer or singers may lock into a particular refrain and render it over and over again, allowing the passions of the performance to pound and pulsate, to swell and strengthen, sending every needle on the WELC control board to its highest register:

> I'm free from sin,
>   and I'm born again.
> I'm free from sin,
>   and I'm born again.
> I'm free from sin,
>   and I'm born again!
> The place where I'm going
> Is out of this world.
> The place where I'm going
> Is out of this world.
> The place where I'm going
> Is out of this world!

Between songs, individual members of the group will burst into testimonials, relating often to the sentiments of the hymn just completed. Sister Dora Justice and Brother Tom Christian are particularly expressive in this regard, contributing to the emotional build-up that leads eventually to Blankenship's sermon. During this process the

178

group shares a sense of timing, playing off each other's contributions and seeming to know in common when upward surges in intensity should be made. By this method the creation of the various moods of the broadcast becomes a collective process, with no individual precisely in the lead. Only during the preaching part of the program does Blankenship emerge sharply in the forefront.

Sister Brenda Blankenship outside the WELC Studio

I mentioned earlier some of the other evangelistic activities that round out Sister Brenda Blankenship's religious involvements: tent meetings, church revivals, street preaching, home services, and performances—with Sister Justice—as a gospel singer. The home services rise out of a strongly established practice in Appalachian Holiness-Pentecostal traditions of worshiping in the home, accompanied by a preacher and some friends and neighbors. The reader will recall that Brother Johnny Ward's "conversion" came about as the result of such a home meeting.

"People just invite me over," says Sister Blankenship. "We don't do nothing special, just sing, pray, testify, and shout—glorifying the Lord. Usually it's a week night, and I do more of this in the spring and summer. It's kind of social, too. But mainly it's for Jesus."[32]

Blankenship's street preaching also occurs primarily during the spring and summer, and these events apparently take place without any prearrangements, except those made by Blankenship, Justice, and Christian.

"We get together in a clear spot in Iaeger, or Bradshaw, or Roderfield and just start singing," says Blankenship. "Usually it won't take too long for some folks to gather. After that we just do what the Lord leads us to do. Sometimes we get to someone who's been backsliding."[33]

Blankenship is often included in the evangelistic activities of Brother Bill Daniel of Kernersville, North Carolina, a Pentecostal preacher who spends his summers conducting tent revivals in Virginia, West Virginia, North Carolina, and elsewhere, traveling with a vehicular entourage composed of a large recreational motor home, a tractor-trailer rig, and occasionally one car. Inside the trailer Daniel transports his four-hundred-person revival tent, a smaller tent (under which are marketed videotapes, audio recordings, gospel albums, Bibles, and a number of other "inspirational" print items), several hundred metal folding chairs, all of his lighting and sound equipment, a large folding performance platform, a number of bulky pieces of musical equipment, all of the necessary tent riggings, signs galore, two large potted philodendrons (for the performance platform), and a generator that provides electricity for the lights and the sound system, the musical instruments, and all the flood lamps used under and around the tent. Two portable toilets also are always needed, but these apparently are secured from local equipment rental agencies.

As was suggested in chapter 2, the message I received from both Johnny and Dewey Ward was that such a tent became a kind of status symbol, an icon that suggested full engagement in the search for lost souls. Sister Brenda Blankenship includes much the same vision in her career aspirations, except on a smaller scale. As indicated earlier, she thinks a two-hundred-person tent would suffice, and she says the tractor-trailer rig would not be necessary—perhaps a smaller truck. "Jimmy could drive it," she adds.[34]

For Sister Blankenship, however, that hoped-for tent means more that just an expansion of her evangelical activity: it constitutes, remember, one of the environments in which she most wants to preach.

Evangelist Bill Daniel calls his travels with his tent the "Miracle Crusades" and his organization the "Bible Revival Gospel Ministries, Incorporated." His traditional mode of operation is to enlist, wherever he stops, the help and support of local evangelistic preachers, particularly Holiness-Pentecostal and Freewill Baptists; and when he comes to McDowell County, West Virginia, for his annual two-week stay, that list of local supporters includes Sister Brenda Blankenship, who in turn uses her radio broadcast to promote the tent meeting. In reciprocity, Daniel works Blankenship into his extensive network of evangelistic connections, helping her secure revival bookings in a multistate area.

During the 1990–91 fall-winter season, Blankenship even preached one revival at Daniel's home-base church in Kernersville, North Carolina. By doing the same for other preachers, Daniel obtains a great deal of local help relative to special music, song-leading, prayers and testimonials, audience building, and convert counseling at the close of services. When I witnessed Daniel's revival in Iaeger, West Virginia, on July 31, 1991, he had the assistance of four local preachers, three men and Brenda Blankenship.

The Tent Revival Service at Iaeger, West Virginia

Iaeger lies fifteen miles west of Welch on Highway 52, one of those numerous old mining camps in McDowell County. Sister Brenda

Blankenship had told me to watch for the large green and white tent that would be on my right just before entering Iaeger's small commercial district. The tent was there, resting on a two-acre stretch of cleared and leveled land, one of the few noncommercial or nonindustrial spaces I had seen in the county that would accommodate such a stretch of nylon and provide all the requisite parking.

This spot obviously was not a natural mountain meadow. Instead, it gave every indication of having been leveled in a strip-mining operation or a bulldozing process that had provided a base for some early mine-related industrial facility. The ground cover contained a heavy mixture of slate and other dug-from-the-mountain mineral matter, and an unhealthy looking, rust-colored run-off had collected in a number of small depressions that lay on the north side of the large tent. Daniel and his assistants may have been forced to fill some of these depressions to make space for this large tent.

Supported by three primary masts, eight secondary poles, and twenty-four peripheral ones—plus a network of nylon lines that ran downward from the three peaks of the rigging, undergirding the sags in the top's fabric—this tent spread out over the approximately 1,800 square feet of packed dirt and crushed rock that constituted the worship space of Evangelist Bill Daniel's Miracle Crusade. Indeed, this was what Daniel called "miracle ground," a consecrated spot, his ready-for-salvation arena—straw-sanctified and canopied for "glory"— a floodlighted oval of space that became both iconic and metaphoric, a clear signal to all who saw the unfurled and soaring nylon that this was ground now specially dedicated. In this instance that "miracle ground" lay in West Virginia, but the locale could have been Virginia, North Carolina, Tennessee, Kentucky, or elsewhere.

The tent's side panels—except those directly behind the performance platform—were rolled up, permitting the night's cool air to fan across the worship space, but also allowing—and this may have been more important—all those full-bodied sounds of revivalism to flow outward along the narrow Tug Fork valley, occasionally catching the ears of motorists who traveled Highway 52, perhaps drawing them to that night's service or to a later one. As an elderly North Carolinian told me once, "The trick is to sing them in." On this late-July evening, there may have been little else in Iaeger, West Virginia, to attract the curious or individuals just looking for people and public activity.

That day I arrived in Welch early enough to drive out to Iaeger in search of this tent, not wanting to chance missing it after the valley had fallen into shadow. Finding the huge green and white spread of

The crusade's tent, Iaeger, West Virginia.

nylon ended up being no problem, and when I sighted the structure I decided to park my car and examine the entire setting more closely, making some notes on the general environment before the scene filled with people.

I strolled up to the tent and noticed a stocky man in his mid-thirties checking out the arrangement of the folding chairs. This was Timmy Nipper of Kernersville, North Carolina, an assistant to Evangelist Bill Daniel.

On first glance, one is not likely to see Nipper as belonging to this setting. Heavy-torsoed and muscular, dark-haired and facially rough cut, Nipper presents a first impression suggestive of a roadhouse bouncer, a person who physically would be able to handle any overly raucous imbiber. In contrast to my stereotyping initial reaction, however, I found Nipper disarmingly polite and cordial, welcoming me to the tent and asking how he might be of help.

I introduced myself, gave him a business card, and told him why I was there, mentioning Sister Blankenship and my interest in her ministry. In turn, he asked me to have a seat and talk a while, observing that he had the tent ready for the evening event and that he had nothing to do but spend several hours guarding all the electronic equipment. In addition to Nipper, two women were busy in the smaller tent arranging the books and tapes that would later be marketed to the revival-goers. It was four o'clock in the afternoon, and the evening's service would not begin until seven. I pulled out a note pad and began to ask Brother Nipper about himself.

Nipper had been with Daniel's "Miracle Crusade" for only a year, having been "saved" under this same tent at a Winston-Salem, North Carolina, revival. It was shortly after that event that he had joined Bill Daniel's enterprise.

The story he told initially sounded almost too proverbially "testimonial," and my first reaction was to wonder how much hyperbole was present in the account. Nevertheless, Nipper's directness and lack of melodramatic tone began to allay my suspicions, ultimately leaving me with the feeling that he had been what he said he had been.

"We were bikers," declared Nipper, "a pretty rough lot. My wife even once rode with the Hell's Angels."[35]

Nipper went on to claim that prior to their "redemptions" he and his wife had devoted several years to a dissolute sort of existence— using drugs, riding with two biker groups, and engaging in other activities at which he only hinted. He told me he had heard Bill Daniel on the radio in the Raleigh-Durham area and had then sought out the Winston-Salem revival. His life, he said, had "reached bottom" and he wanted a change.

Today Nipper spends his spring and summer months traveling with Daniel's crusades, helping with the tent set-ups, testifying to revival audiences, leading nightly prayers, collecting offerings, assisting in altar calls, and occasionally leading the singing. His days are spent cleaning the tent arena and surrounding grounds, laying down fresh straw, straightening chairs, checking all electrical equipment, making minor repairs in the rigging, and watching over the entire facility.[36]

After Iaeger, West Virginia, the crusade was scheduled to move to Galax, Virginia, for a three-week stay. That would mean all the necessary dismantling, repacking, transporting, and reassembling, tasks in which Nipper's strength would be valuable. He and Bill Daniel were the only males in the crusade's traveling entourage, but help frequently came from local preachers and their congregations.

Although I was not exposed that evening to one of Timmy Nipper's testimonies, I had no trouble imagining how his story might have been received by two or three hundred supercharged revival-goers. Here was a man who had experienced some of the worst of "sins"—drugs, sexual license, the biker's life, cynical disdain for the faithful, violations of law—and perhaps more and worse. Now, however, "Jesus had set him on a new course," and he stood as an example of what "redemption" could do for all men and women. Such a testimony would mesh well with the emotionally charged evangelistic rhetoric Brother Bill Daniel would employ in his nightly exhortations to salvation.

While talking to Nipper, I had a chance to study the complete setup, a collection of revival-crusade gear that would have made Johnny Ward envious. Under the tent, rows of folding chairs extended along one long side and around the two ends, semicircling the large platform (twenty feet wide and sixteen feet deep) that constituted the main performance area, except when Daniel or other preachers and testifiers surged out into the audience. Access to this platform was provided by a centered smaller riser with ramps that at right and left ran down to the straw-covered ground.

On the platform rested a number of microphones (four in all), three tiers of speakers (one tier for the organ and the others for additional musical instruments and the mikes), a large amplifying unit, an elaborate drum set, a small collection of electrified stringed instruments (each resting in a foldable metal stand), a large Hammond electronic keyboard (sitting on a curtain-fronted support frame), and the two potted philodendrons (flanking the platform at right and left). A considerable financial investment in evangelism could be seen in all this musical equipment and other platform paraphernalia. The dominant image, however, was one related to theatricalism, show business, and musical entertainment.

A twenty-five-foot banner topped this picture, proclaiming the biblical message "No man can do these miracles . . . except God be with him," followed by a drawing of a Bible with "BRG Ministries" superimposed across its opened pages. Finally, an expansive American flag hung behind the platform's center section, mixing a symbol of patriotism with the space's numerous religious icons.

Immediately in front of the platform the hard ground was covered by a section of indoor-outdoor carpet, approximately twenty-five feet wide by fifteen feet deep. It was in this carpeted area that most of the revival's emotional drama took place that evening. This was the "pit," serving the same purpose as that served by the similar area in Brother Dean Fields's Thornton Freewill Baptist Church.

Hanging above this worship space, clamped to the tops of the three center tent masts, there was a long electrical batten, from which hung a number of lighting instruments that served to flood the region below with high-intensity illumination. All of the electricity required for these lights and the platform's various pieces of equipment was supplied by the portable gas-powered generator that sat somewhere in the trailer parked behind the tent. At the beginning of an evening service this generator can be heard, but the exuberant sounds of revival soon overpower the equipment's constant putt-putting sound.

The smaller tent was set up at the west end of the larger one. Under this auxiliary facility one or two female representatives of the "Bible Revival Gospel Ministries, Incorporated" were always marketing (except during the actual service) a wide array of religious items— Bibles, gospel recordings, videotapes of other revival services, Bible-story coloring books, various inspirational reading materials, and a number of decorative religious artifacts. Among these items was one videotape of a revival service conducted in Brother Bill Daniel's Kernersville, North Carolina, church—a service at which Sister Brenda Blankenship was the visiting evangelist.

The evening's revival was scheduled to begin at seven o'clock, but some worshipers began arriving at six, driving into the parking area in cars, vans, or pickups, and then often remaining in their vehicles for twenty or thirty minutes before entering the tent. One late-model van contained six adults who had stopped by a fast-food restaurant on the way and who now enjoyed a meal of burgers, fries, and soft drinks prior to the singing, testifying, preaching, and healing that would constitute the central attractions of their evening.

During these arrivals I began to notice an interesting phenomenon: women would emerge from vehicles and come under the tent, leaving men sitting in their cars or trucks. Indeed, these men often remained in their vehicles for the duration of the evening—windows rolled down, listening but emotionally disengaged. Throughout the revival service, even those men who had initially entered the tent would return to their cars, sometimes to smoke, but at other times merely to break connection with the emotional activities of the evening. Their wives sat with other women under the tent and participated vigorously in the singing, shouting, and praising. Most of the arriving vehicles contained only women.

At the peak of the night's attendance, approximately 200 worshipers were present, around 150 of whom were female. In addition, at the close of the evening, when numerous participants came to the pit for laying-on-of-hands experiences (soon to be described), the clusters were composed solely of women, except for the male preachers involved with the service. In several ways, this revival became a female affair, particularly when one considers only those who played respondent roles, as opposed to those who assumed leadership roles. It was always the female voices—through singing, shouting, praising, and testifying—that provided the sound surges that in turn generated more singing, shouting, praising, and testifying. Aside from Daniel, Nipper, and one other male who led much of the singing, this event became a feminine religious expression, with Sister Brenda Blankenship and Sister Dora Justice featured prominently in that expression.

The service started with song, those high-volume, fast-paced, arm-waving, feet-stamping, shouting-accompanied, joyous hymns that characterize the Holiness-Pentecostal worship style. Led by Blankenship and the one male song leader, this happy but unsophisticated sound flooded the tent, poured out along Highway 52, and surely must have reached the several homes that dotted that length of the valley, perhaps catching the attention of individuals who had sought the cool air of front porches, and—just maybe—"singing 'em in." In truth, a number of worshipers did arrive long after the singing had begun.

This was music that required people to stand, and sometimes to

Bill Daniel, *on the drums,* with his wife, *at the organ,* and Timmy Nipper, *at far right.* A local preacher leads the singing.

move about—to dance in aisles, to hug other communicants, to twirl in kinetic expressions of reverie, or simply to wave extended arms. However, these were not hymns requiring subtle harmonics, fine vocal skills, or a songster's appreciation for poetic forms. Instead, these were musical works that cried for passionate expression and full-throated projections, singing that was accompanied by all those previously noted amplified musical instruments, singing that often approached shouting.

Early in the singing service, Sisters Brenda Blankenship and Dora Justice—along with a third woman whom I did not identify—assumed positions on the platform and brought the audience to their feet with several particularly lively hymns. Both Blankenship and Justice led in the arm-waving physicality of this worship in song, occasionally emitting shouts that sparked new levels of audience exuberance.

For a while Bill Daniel played an electric guitar, and then switched to his drums, always providing a model for the involvement he expected from his audience. His wife operated the crusade's electric organ, sensing every instance of need for a surging sound. Indeed, Sister Daniel's organ was particularly effective in arousing audience response. Not only during the singing, but also during preaching and testifying, her dramatic and spirited chords intensified whatever might be the expression of the moment—swelling during passionate hymn refrains, dramatically supporting all rhetorical contributions of note, and providing emotional background for any come-for-conversion-or-healing action that might transpire in the pit. So pervasive was this organ sound, in fact, that one tended to forget it, sensing only the totality of the particular moment of expression.

The effect of the organ music became especially noticeable, however, at the close of Daniel's sermon. Throughout the earlier parts of that sermon the evangelist's wife was away from her instrument, having positioned herself behind a tier of large speakers (out of sight to the audience) where she spent those moments sorting and counting the offering receipts for the evening. Then, as if by a husband/wife understanding of long tenure, she moved back to the organ and struck a cord or two. In turn, Daniel intensified his delivery and the two began to work in unison toward a sermonic climax, carefully measuring the moment's moods to avoid an ineptly rushed ascension.

During these moments it was not just the preacher and organist who were locked in these mutual response patterns: The audience was part of an interactive dynamic process, frequently being moved

Sisters Brenda Blankenship, *right,* and Dora Justice, *center.*

but also becoming a mover—willingly yielding itself to an organic interplay in which the roles of stimulator and respondent fluctuated between two or more agents. The audience had come to this tent prepared to be part of an emotional and spiritual fermentation, one for which they were an essential—and willing—element. An understanding of that reality helps explain events that transpired at the close of this sermon.

When it became obvious that Daniel's sermon was reaching its climax, a number of women in the audience quickly expanded their contributions to the rapidly escalating dramatics of the moment—standing and clapping, moving outward into aisles, executing the standard arm-waving maneuvers, and shouting and jumping. One woman immediately rushed to the pit, threw her arms in the air, and began to twirl in ecstatic jubilation, action that she would continue until she collapsed on the waiting carpet, the first of many females who would "swoon in the Spirit," also referred to as being "slain in the Spirit" or "dying in the Spirit."[37] Other women began to crowd forward, without being called, to have Bill Daniel lay his hands on their foreheads or temples. In most cases such an act would result in the woman's swooning to the ground, always being eased gently backward by other female worshipers or by one of the male assistants.

Whenever a woman—and it was always a woman—fell backwards in one of these spiritual faints she would maintain the precise body position she had held at the moment of swooning, frozen in her granitelike pose, without any movement, for as long as she lay on the carpeted ground. That meant that if she had had her arms in a particular configuration at the time of collapsing, she would maintain that same posture during the entire duration of her swoon, sometimes as long as ten minutes, even when her arms were in the air—away from her body—in what could become a highly fatiguing position.

One of the women who had earlier managed the marketing tent now stood prepared for the scene developing in the pit: As a participant fell backward to the carpet, sometimes exposing more of her legs than was seemly, this woman decorously draped the swooner's lower extremities with a small velveteen coverlet, reclaiming the fabric once the raptured female had revived. In fact, there existed within this sisterhood of worshipers an ethic that required full attentiveness to the needs of enraptured participants—supportive actions ranging from the initial breaking of a fall, to subsequent rearrangements of disordered clothing, to assistance when the revived reveler sought to stand. During the twenty to thirty minutes devoted to this "swooning

Swooning in the spirit.

in the Spirit," I saw one pair of women who, in full reciprocity, reversed roles for each other, the swoon and then the assist, or vice versa. The respective episodes were not so tightly juxtaposed as to suggest plan, but the commitment to a shared attentiveness was obvious.

One favorite response to these final moments of revivalistic fervor was for several women to cluster around either Bill Daniel or Brenda Blankenship for what would become a group "laying on of hands," either collectively concentrating on one worshiper or switching roles so that each received the blessings of touch. These episodes were not healing sessions, or at least not overtly so. Instead they served merely to engender in each recipient a higher level of spiritual experience. Some of these clusterings resulted in one or more of the women swooning, while others did not, becoming merely impassioned exchanges of a mutually experienced religious excitement.

What became obvious throughout this service—particularly relative to women—was that the emotional highs obtained depended heavily upon a group process, the energy dynamics that arise only from the interactive contributions of several participants. Probably no one of these female worshipers would have been able to generate her evening's personal high point in emotional expressiveness without stimulation from the group. Crowd contagion was at work, but often its fullest manifestation was found within these clusters of six to ten enthusiasts.

For quite a while I saw no evidence of a conventional "coming to redemption" breakthrough. Most of the participants who were active in the pit appeared to be involved in intensification experiences: They already considered themselves "saved" and were participating in joyful "praising" and "glorifying." Furthermore, Brother Bill Daniel had never issued any formal "call to redemption." Instead, the down-front actions had developed purely in response to an understanding that the time had come when individual audience members could be fully expressive of their own religious zeal, free to act rather than just react.

A different set of actions developed, however, when at one point Sister Brenda Blankenship suddenly rushed out into the audience to speak privately to one audience member, a dark-haired woman, perhaps in her mid-thirties, who was sitting among a small cluster of female friends with whom she had come to the revival. The drama that followed was so rapid in development that it gave the appearance of lacking precise motivation: The actions were explosive and without

Swooning in the spirit, and the coverlets in place.

Clustering in the pit. Blankenship *at right center.*

any connection to a tension-building chain of happenings that I had been able to observe.

Almost immediately after Blankenship reached the woman in question the latter jumped to her feet and began shouting and crying, loudly enough that her behavior took focus away from the myriad other dramas transpiring under this tent. She subsequently allowed herself to be led forward to the pit, where she collapsed in abject prostration on one of the ramps leading to the performance platform. There she lay and sobbed uncontrollably, her body racked by spasmodic jerks and deep shudders. Blankenship sat nearby and cried herself.

Finally Sister Blankenship jumped to her feet and grabbed a microphone from its stand on the platform. "I want you folks to hear some happy cries of salvation," she said, positioning the mike in the sobbing woman's face. Immediately the woman emitted a prolonged cry, to which the audience responded with applause and shouts.

Twice the woman tried to rise, falling each time to her knees for more unrestrained weeping, now having to be supported on one side by Blankenship and on the other by Timmy Nipper. Indeed, this woman was totally incapacitated, and I do not think I have ever seen anyone cry as forcefully as she did, nor perhaps as long. For ten to twelve minutes this woman poured forth a convulsive form of grief and joy, purging herself of some intense sense of guilt, shame, remorse, or what have you, never saying anything that I could hear that would explain her turmoil. Watching her was absolutely exhausting, and perhaps a little frightening. The emotional and physical controls that make us feel safe around people were gone. For whatever reason, this woman had yielded herself to a total release of tensions, and the resulting kinetic display seemed on the verge of frenzy.

Finally she succeeded in rising and moving back to one of the front-row chairs, where Blankenship joined her for additional consoling or counseling. The other women who had come with her that evening now moved forward, forming a feminine support system around the mourner as she gradually settled into a calmness of spirit that suggested catharsis.

I had no opportunity to interview this individual, as I had done with the woman in Drexel, North Carolina; but I confess that I was very curious about motivations, so sudden and so unexplained had been the explosion. That curiosity has never been completely satisfied, but on August 12, 1991—when I was back in Boone, North Carolina—I called Sister Brenda Blankenship, needing more details

Daniel and Nipper working the pit clusters.

about her own life. During the phone conversation that ensued, I mentioned the episode under the tent and asked why she had, in the first instance, singled out this woman for an approach.

"I knew her," she said, "and I knew her to be a backslider. She had been with the Lord, but now she wasn't living right. So when I looked and saw her, and saw her face telling me something, I decided to go to her. God called me to do that, you know."[38]

Evangelist Bill Daniel Raises Money for His "Crusades"

I indicated in chapter 1 that eventually I would be detailing two situations involving money solicitation. The first was my discussion in chapter 2 of Brother Johnny Ward's comparatively low-key campaign to secure a tent, and the second will be an examination of Evangelist Bill Daniel's much more high-pressured appeal for money.

Daniel's "Miracle Crusades" necessitate a heavy investment in equipment. As already indicated, this equipment inventory includes a tractor-trailer (a relatively new rig), a large motorized recreational vehicle (apparently furnished with a kitchen/dining/living area and sleeping accommodations for at least four), the two tents and all their riggings, extensive lighting and sound equipment, numerous musical instruments (including the large Hammond electronic keyboard), the network of risers that constitute the performance platform, hundreds of folding chairs, a great variety of promotional signs and banners, and all that stock of items marketed under the smaller tent. During that July evening in Iaeger, West Virginia, Daniel mentioned all this equipment only once, and that was during his call for contributions. At that time he said simply that he hoped to finish paying for the truck and trailer that summer, perhaps during the Galax, Virginia, revival. This was all "the Lord's equipment," he said, for which the people would be "helping him pay."

This "helping him pay" scene transpired between the singing service and Daniel's sermon. As Sister Daniel played her organ, and as Nipper and one other male assistant stood ready with numerous envelopes, Daniel announced that he would be asking people to donate specific dollar amounts. "This first group of envelopes," he declared, "are one thousand dollars envelopes. Who will take one?"

After considerable urging, Daniel actually got one individual—one of the few men in the audience—to accept a thousand-dollar envelope. The man in question happened to have been a person with whom I had talked prior to the service. Among other things, he told me that he was familiar with the preaching of Bill Daniel, having attended one of the evangelist's crusades in Virginia. There's always the possibility that this man was a "plant"; however, I have no specific reason to believe this to have been the case.

Next, Daniel moved to the five-hundred-dollar envelopes, but after several minutes of entreatment he still had not convinced anyone in the crowd to accept one from this category; therefore, he moved on to the one-hundred-dollar envelopes, at which time he obtained success with a half-dozen or more worshipers.

Daniel then dropped successively to the fifty-dollar contributions (some takers here), the twenty-dollar ones (far more this time), and finally the general collection, for which the traditional offering plates were passed. All of these pleas for money consumed approximately twenty minutes, and I hesitate to estimate how much money was given during this period. If all the envelopes that were accepted were returned with the designated contributions, then these two hundred or so West Virginians became particularly generous, given the obvious economic condition of McDowell County. The two-week stay in Iaeger certainly must have produced some profit for the Bible Revival Gospel Ministries, Incorporated.

Concluding Thoughts

I will not close this final case study with the focus on Evangelist Bill Daniel's money-raising techniques. Instead, I am returning to a topic that has served as an integrative theme in this chapter: the role of women in these McDowell County, West Virginia, religious happenings.

Although Bill Daniel preached the sermon that July 1991 evening, there was much about the larger event that made male involvement only peripheral to all that transpired under that tent. First, male presence in the audience was marginal, and those men who were there gave little to the evening's total dynamism. The passion that imbued all the singing, shouting, crying, dancing, and touching was predominantly female-driven; and, aside from the presence of Bill Daniel, Timmy Nipper, and one local male preacher, the pit became a female preserve. Without the energy spent by women in that carpeted area—and throughout the audience—the events of the evening would have been relatively tame. Indeed the least energetic parts of the service transpired during Daniel's own sermon, at moments when he was making no specific requests of the audience, such as "Let's hear someone amen that thought."

Second, the strongest leader of the evening turned out to be Sister Brenda Blankenship. I had expected Daniel to play a far more forceful role than he did, orchestrating the various emotional demonstrations with the skills of an evangelical preacher of long experience. Instead, he seemed to recognize that these were largely Blankenship's people and that this Sister possessed a local influence rife with possible benefits for the crusade. Even Daniel's sermon seemed subdued in comparison to the explosive exhortations I had heard Blankenship provide during her radio broadcasts.

Another factor that made Blankenship stand out for me was that Timmy Nipper did not live up to my earlier expectations. After hearing his story about his life with the bikers, I thought his testimonial would find a place in the evening's drama. As I suggested earlier, I visualized how such a narrative might play with an already electrified revival audience. Even now, my suspicion is that Nipper had made his testimonial contribution earlier in the week. His involvements the night I attended were largely of a background nature.

Finally, these McDowell County women seemed not to need a great deal of what I once heard a Missionary Baptist call "the preacher's starter fire."[39] At the close of Daniel's sermon he needed only to provide the appropriate signals that let these female worshipers know the stage was theirs. In response, they filled the pit with remarkable speed, so much so that I was left with the impression that at least some of those women had been impatiently waiting, thinking to themselves, "When's he gonna let us get really involved?"

I will advance one speculative thesis: There may be in McDowell County, West Virginia, a socioeconomic environment that fosters female empowerment—in religion, politics, or what have you. Male population figures have fallen faster than those for females; and there is some evidence that young males in particular flee the county, in search of greater opportunity in other markets of commerce and industry. A comparatively generous state welfare system, on the other hand, allows single female parents to support households under conditions that might be less appealing to a male.

As I have already shown, fifty years of population changes in McDowell County have produced a male/female distribution in reverse of what it was during the region's boom years. Old-timers tell of boarding houses up and down the valley crammed with single men who worked the mines. Indeed, one man told me of a facility that became so crowded it slept men in shifts, two men assigned to each bed but occupying it at different hours.[40] Rex and Eleanor Parker, who entertained at many of the county's theaters, schools, and union halls during World War II, say their audiences frequently were heavily male, and those were years in which much of the nation's male population was in the military.[41]

Today young men find it difficult to find steady employment in the county. According to Mountain View High School (Welch) social studies teacher William L. Kell, Jr., "They leave the county as soon as they graduate, and most don't return."[42] Ralph Smith, a Welch studio photographer who is retiring, and at the time of this writing trying to find someone to purchase his business, told me that the civic club to which he has belonged for years no longer has any young members. "Our most junior member," he said, "is well into his fifties. That's kind of the nature of this town right now."[43]

None of these speculations is what I would call "baseline-critical" to this case study of Sister Brenda Blankenship and her activities related to the airwaves of Zion, but they may help explain why the Iaeger, West Virginia, tent revival became such a female-dominated happening, even after being initially led by a male. Finally, women of McDowell County—and perhaps elsewhere in Appalachia—undoubtedly find some comfort in the Holiness-Pentecostal codes that allow them to take charge of their religious institutions and practices, even when such actions are not necessarily mandated by an absolute dearth of males. Readers familiar with Appalachian "old-time" Baptist subdenominations will recognize that such gender-role shifts are not at all easy in those religious environments. In Appalachian Holiness-Pentecostal settings, however, Blankenship finds plenty of support for her acceptance of Joel's proclamation "and your daughters shall prophesy."

—6—

Some Common Threads

We believe the Lord called us to his stage. It's not the Grand Ole
Opry, but it's certainly grand.

Eleanor Parker

Loyal Jones warned scholars against easy generalizations relative to
Appalachian religion,[1] and I feel compelled to communicate the same
message, urging readers to avoid sweeping judgments drawn merely
from the four case studies in this volume. There are aspects of these
cases that make them representative, as indicated in chapter 1, but
there are also numerous airwaves-of-Zion individuals and groups—
in addition to those mentioned—who possess their own distinctive
characteristics. The phenomenon is multifaceted, rich in the same
cultural diversity found in the larger Appalachian scene. As Jones
has said, "Appalachians, like every other group, are a varied lot even
if they have certain characteristics in common."[2]

We have these four case studies in front of us, however, and since
we do I want to examine some of the characteristics they share in common. Are there any threads that run through the fabrics of all four?
Yes there are. I will start with the concept of *mission*.

Sense of Mission

One of the strongest elements in the broader Appalachian religious tradition is the "call" thesis, the idea that God summons the be-
liever to all factors of his or her spiritual life. The more Calvinistic of
the region's religious groups see this "call" as merely another part of a
larger predestination scheme, while the more Arminian groups view
"call" as an extension of God's involvement with the individual, a be-

lief that the deity addresses needs of individual humans in very discrete ways.

"Call" in turn becomes the foundation for "mission," what Old Regular Baptists usually call "gift":[3] a combination both of life direction and the amalgam of skills, talents, personality characteristics, and circumstances that allow the following of that direction. "Mission," therefore, is not just what you want to do: It is what God wants you to do and has equipped you to do. The abandonment of "mission" is more than violation of self: It is violation of God.

In all four of the airwaves-of-Zion cases we have examined, this factor of "mission" comes into play. The central personae in each instance feel called to their particular religious involvements, and in each instance this sense of call has produced a sense of duty.

With very specific language Johnny Ward declares that he has been "called to prophesy," and he dates that call from the time of his conversion at his sister's house in Mount Pleasant, North Carolina. Furthermore, it is revealing that Ward employs the verb "to prophesy," rather than "to preach," "to evangelize," or some other option. "To prophesy" carries that Old Testament imagery of the itinerant spiritual seer, wandering a land to proclaim a message of doom for those who do not turn from error. Thus the term blends well with Ward's dream for a tent and a converted school bus, the requirements for his own hoped-for journeys through the land. The reader will also recall that Johnny Ward claims to have been told by God that he will produce another "prophet," and he thinks that individual will be his son, Wesley.

I will not, however, place too much emphasis on the tent part of this man's vision, for Johnny Ward views all of his individual evangelistic activities as belonging to this called mission, including his radio program and his more traditional church revival meetings. He declares, as does his wife, Sadie, that all his work is God-directed. Furthermore, it should be remembered that in pursuit of all these mission activities he has shown himself willing to forgo the securities of regular employment, placing himself (according to his declaration) completely "in the hands of God."

Rex and Eleanor Parker believe their real mission began in 1959. According to them, the years from 1941 to 1959 (and even earlier for Rex) had been preparation for their respective calls to the world of spiritual entertainment, providing the talents (gifts) that would then become redirected—their transformations from "hillbillies" to "holybillies." "We believe the Lord called us to His stage," declares Eleanor. "It's not the Grand Ole Opry, but it's certainly grand."[4]

One factor making the Parkers' sense of mission so interesting, however, is the commercial broadcasting element that has remained in their particular programming. As was suggested earlier, Eleanor weaves the needs of her sponsors—and the services they supply to a deserving public—into the total fabric of the Parkers' "Songs of Salvation" mission. The assumption is that he or she who buys a chest of drawers from Gayle's Resale joins forces with that mission.

Brother Dean Fields dates the beginning of his mission as 1982, the year he turned his life around and founded the Thornton Freewill Baptist Church. The personal associations he subsequently established—with the scores of individuals who serve both his church and his radio broadcast—helped him, he says, as much as he helped others. At the moment he sees his life as being firmly fixed to God's plan, spreading "words of love" along Thornton Creek and throughout much of Letcher County.

In contrast to Johnny Ward and Brenda Blankenship, Dean Fields places emphasis on his role as "pastor," envisioning no tent that would take him away from the congregation he has nurtured and that in turn has nurtured him. Recall his statement "I'm a pastor, not a master." The declaration suggests a personal mission strongly tied to an interactive ministry, one in which the preacher is only part of the total dynamism that sparks these congregational gatherings on Thornton Creek. Fields, therefore, sees his divinely decreed role being played out with the particular population that constitutes, or will in the future constitute, the Thornton Freewill Baptist Church. His "Words of Love" broadcast is merely an extension of that involvement.

Like Johnny Ward, Sister Brenda Blankenship sees her mission as an evangelist, a "gospel peddler."[5] As I noted several times, she claims to become more intensely inspired when she preaches outdoors or under a tent; although it is difficult for me to imagine an exhortation intensity more elevated than that present in her typical radio sermon. I must quickly note, however, that I have never seen her in action during one of her street-preaching episodes, an activity that brings her, she says, a high sense of fulfillment.

When I first met Blankenship I asked her—as I am accustomed to doing—how she would like for me to refer to her. First she said "Sister Blankenship," and then she said, with a bit more assertiveness, "Evangelist Brenda Blankenship," the latter suggesting not only a name and title but a promotional label, the words one might find on an advertising leaflet, a roadside sign, an across-an-entranceway banner, or a side of a truck or trailer. Sister Blankenship clearly sees her mission as involving some mobility. A cordless mike and a traveling ministry: these appear to be critical factors in this woman's aspirations and in her perception of what God wants her to do.

Their Missions Arose out of Troubles

A second factor present in all four of these four case studies is that missions—or the sense thereof—have arisen out of a bout of personal troubles. In each instance the individual or individuals adopted the respective spiritual path after a period of personal disturbances—some physical, some psychological—with alcohol figuring in all four stories.

Johnny and Sadie Ward assumed their current course after a serious automobile accident (involving both Johnny and Sadie) and an alcohol problem (involving only Johnny). Rex and Eleanor Parker took on their "Songs of Salvation" mission after a session of deep depression (Eleanor) and, again, a developing alcohol dependency (Rex). Drinking also had been one of Dean Fields's problems, in addition to his 1981–82 series of cancer surgeries. Finally, Brenda Blankenship suffered the trauma of three deaths in quick succession—her father, her sister, and her daughter, the latter as a consequence of drinking and driving on the part of her son-in-law.

Turning to religion as a response to one's problems always engenders concerns about spirituality as a crutch—Marx's "opiate of the common man" principle. In fact, Loyal Jones has argued that observers of Appalachian religious traditions have been all too eager to use the crutch metaphor when examining motivations of mountain people, while not being as eager to do so when discussing practices in more mainline faiths. "That which is a crutch to mountain people," says Jones, "is a solace to others, it seems."[6]

The central characters of these four case studies probably would reject both terms ("crutch" and "solace") when explaining their personal motivations for these life changes. Instead, they would argue that God used their respective travails as part of his way of "calling," jarring each of them out of their particular states of misdirection or apathy. Suffice it to say that the Wards, the Parkers, Dean Fields, and Brenda Blankenship view the troubled times through which they moved as passages to new life-styles that are in God's will. Were other career changes—of a spiritual nature—to develop, those would also be in God's will.

The Passion to Communicate and Specific Messages

A passion to communicate is present, to some degree, in all humans, but this drive appears particularly strong in the key people of these case studies. In one way or another, each told me of his or her strong desire to send forth specific messages, in these particular instances through the medium of the airwaves of Zion. My intent here is not to equate this drive with ego; instead, I am more comfortable relating it back to "call" and "mission," simply because that's the way the persons involved would explain the compelling force they say is operative.

Regardless of the motivating impulse, the passion is there, and it pushes these individuals to be public people, at least within their highly restricted worlds. They seek opportunities to be heard, to influence, to persuade; and in each case this drive, process, and consequence is tied to a personal message.

For Dean Fields this message is embodied in the title of his broadcast, "The Words of Love." As indicated earlier, Fields imbues his sermons and his behavior with that strong commitment to "Christian love." The concept becomes both his theology and his communal code, equipping him to work with people who often have become unlovable to others. It also endears him to his congregation to such a degree that a very visible form of group cohesion develops, centered around his personal warmth. I remember well that Saturday evening service during which he suddenly called an elderly man to the pit and told his church that insufficient appreciation had been shown this individual, that everyone should come forward and embrace him. Everyone did just as they were told.

Since Eleanor becomes the primary spokesperson for the Parkers, I need to focus on her when trying to identify the pair's major message. If one attends to Eleanor's religious rhetoric for very long, one will hear a particular thought over and over again, a thought that is expressed in the title of one of her own songs, "He'll Set You Free."[7]

The statement becomes Eleanor's central evangelistic thesis, arising from some deep conviction that what happened for both Rex and her was a liberation. By the late 1950s she had lost much of her earlier zest for performing and had found herself sinking into poorly motivated routines, relying heavily on talent and technique unblessed by a "joyous spirit." Rex may have been experiencing the same problem, turning him to alcohol. They both now depict their "spiritual rebirths" as revitalizations that set them "free."[8]

Eleanor appears to support the idea that there is a point at which the creative soul and the spiritual soul are one and the same, a blending of the muse and the mystical. It's an appealing thesis, and when I heard these two aging performers at Brother James R. Kittinger's church in Pilot, Virginia, I received the impression that Eleanor believed they had reached that blending point. They certainly were happy and in their element.

The dominant message coming from Johnny and Sadie Ward—and also from Dewey Ward—is that their role is to work with a type of person who cannot find a place in the more establishment-oriented mainline faiths. Both Johnny and Dewey told me that their Pentecostal practices were more appealing to, and provided a place for, some folks who would not be comfortable in more "uptown" churches. Emotionalism was mentioned on these occasions, but so were some other things—education, dress, testimonial habits, and basic style of worship.[9]

When making these arguments, Johnny and Dewey Ward do not accept—for themselves, their mission, and their followers—a relegation to inferior status. Instead, they speak as if recognizing that people are different, some needing a particular type of religious expression, and some another. However, Johnny did make it clear that he thought a true religious experience should trigger something dramatic in a person, the event being that important.[10]

The message I want to mention as coming from Brenda Blankenship may be one I have assigned to her, as opposed to her assigning it to herself: her argument for an accepted validity for the female exhorter. I recognize that I needed a spokesperson for this "sisterhood" of airwaves-of-Zion personae discovered at WELC. Blankenship provided me that spokesperson by saying all the right things, by voicing a strong defense for female preaching.

It needs to be recognized, however, that Sister Brenda Blankenship can become quite forceful about this issue. After all, the arguments she must counter declare all of her evangelistic activities to be manifestations of apostasy. That's certainly the declaration she would receive from Appalachia's "old-time" Baptists, and from a sizable number of Southern Baptists. Therefore, Blankenship is fighting for her own validity, her own sense of being in the proper spiritual order of things.

All of this, however, is a nonissue within the Holiness-Pentecostal environment in which Sister Brenda Blankenship operates. Apparently none of the women who crowded the pit during that tent revival at Iaeger, West Virginia, would give the question the proverbial second thought. Indeed, they might become surprised—or even angered—when someone like myself brings the issue into focus. Blankenship's explanation of what has been interpreted as Paul's condemnation of female worship leadership would, no doubt, completely satisfy these women. Furthermore, some of them might say, "We don't really need Brother Bill Daniel. Just give us Sister Brenda."

Spiritual Optimism

In his essay "Churches of the Stationary Poor in Southern Appalachia," Nathan L. Gerrard indicted much of the region's "Holiness" religious expression as being mired in fatalism. Contrasting "stationary poor" with the "upwardly mobile poor," Gerrard argued that these "stationary poor" "Holiness" elements in Appalachia allowed their fatalism to corrupt almost every facet of their social system. The image emerging from his essay is of a people so debased in character that one might not want to call them human. "This [fatalistic] outlook," wrote Gerrard,

> can express itself in two sharply contrasting ways: The first expression is cynical and pessimistic, symbolized by the phrase: "Nothing good will happen to me." *Pessimistic fatalism* is manifest in *squalor*—the failure to make the most of the little one has: dirty dishes in the sink? unrepaired, tattered furniture; litter in the rooms and around the unpainted shack; bugs and sometimes rats. It is also manifest in *reckless hedonism,* the conviction of those who lead insecure, unpredictable lives that a pleasure postponed is a pleasure forever lost. . . . Reckless hedonism results in noisy drinking bouts, illegitimate births, incest, absenteeism or quitting of jobs in order to go hunting or fishing, spur of the moment purchases of luxuries when money is needed for necessities, and other kinds of behavior incomprehensible from a middle class point of view.[11]

While recognizing that some of these human weaknesses occasionally do become "manifest" in the followers of Johnny Ward, Rex and Eleanor Parker, Dean Fields, and Brenda Blankenship (all showing influences from the Appalachian Pentecostal-Holiness tradition), it would be decidedly unfair to suggest any degree of tolerance on their parts for these behavioral aberrations. Furthermore, the declaration of a cause and effect relationship between any Appalachian religious tradition and such social disorder seems highly speculative at best: Urban ghettos of the Northeast and upper Midwest have often become the locales for the same human conditions. Poverty has a way of debasing human behavior wherever one finds it. In addition, I think I have seen a substantial degree of "reckless hedonism" on every university campus at which I have taught and among a number of my "middle-class" friends.

The main point I need to make, however, is that in one sense these airwaves-of-Zion persons are highly optimistic. They believe they are a part of a faith system leading not only to a "glorious" afterlife but to a far better present life. Brother Dean Fields, in particular, would be hurt by a charge that his religious beliefs and practices are conducive to behavioral debaucheries, since that was the very life-style from which he claims to have led a number of his church members. Dewey Ward would feel the same way, arguing that he played a role in moving a Drexel, North Carolina, woman away from some of the evils Gerrard identifies. Brenda Blankenship would point to the woman who came forward under Bill Daniel's tent, and each of these individuals would identify his or her own religious experience as evidence of the civilizing effects of the faith.

Finally, all of these airwaves-of-Zion persons would take issue with Gerrard's fatalism argument, reasoning that their beliefs represent the epitome of optimism. Life is always open to change, they say, and change for the better. All one needs to do is to renounce the old and take on the new. Furthermore, why blame all those evils on the Holiness movement? A significant contribution of the Holiness movement has been the idea that human nature is perfectible. Proponents of this theological position would have difficulty understanding why Gerrard charges the faith with responsibility for such obvious imperfections. It's the Calvinists, they argue, who are more likely to be fatalists.

Final Observations

In chapter 1, I argued that the airwaves-of-Zion phenomenon is fading. I still believe that to be the case; however, all the peripheral activities I have examined in this work (the revivals, etc.) will not pass from the Appalachian scene so quickly, if ever. This regional culture will, nevertheless, lose some of its richness when the remaining airwaves-of-Zion programming goes the way the "Morning Star Gospel Program" has gone. When that occurs, there will be many who will mourn the passing of the phenomenon and the particular "radio-land" contributions made by such individuals as Sister Dollie Shirley, Brother Roscoe Greene, Sister Ann Profitt, Brother James H. Kelly, Sister Myrtle Lester, and Brother Douglas Shaw, plus the multitude of other Appalachians I have mentioned in this volume.

It's become my tradition to close a work by expressing gratitude for the openness shown me during my Appalachian religious studies. I need to do that again, since airwaves-of-Zion groups have been remarkably receptive to my presence in their studios or churches, even when I was taking notes and photographs, and frequently when I was tape-recording. I occasionally have to remind myself that I do not have to endure many visitors in my university classrooms when I am teaching, particularly strangers with cameras, notepads, and an overabundance of curiosity about me as a subject of ethnographic interest.

During the last four years I did find myself turned away from three broadcast studios. In addition I had two groups say yes when I asked them if they would mind my taking pictures. Responses such as these never angered me, because I was well aware of my interloping status. Rejections, however, did make me more appreciative of those Appalachians who received me with the region's traditional openness, particularly those persons connected with the four case studies in this volume.

Notes

1. The Airwaves of Zion: An Overview

1. I am employing this term in the same spirit and usage as David Edwin Harrell, Jr., gave it in "The Evolution of Plain-Folk Religion in the South, 1835–1920," in *Varieties of Southern Religious Experience,* ed. Samuel S. Hill (Baton Rouge: Louisiana State Univ. Press, 1988), 24–51.
2. Drawn from association minutes for various "old-time" Baptist groups, including Regulars, Old Regulars, Uniteds, Separates, and Primitives. The term does not appear to be as popular in church names for some other denominations touched on in this work.
3. *The Cokesbury Worship Hymnal* (Nashville: Abingdon-Cokesbury Press, 1938), hymn no. 75.
4. Additional insight into Holiness and Pentecostal influences upon the South and Appalachia can be gained from Harrell, "The Evolution of Plain Folk Religion in the South," 38–42, and Troy D. Abel, "The Holiness-Pentecostal Experience in Southern Appalachia," in *Religion in Appalachia: Theological, Social, and Psychological Dimensions and Correlates,* ed. John D. Photiadis (Morgantown: West Virginia Univ. Press, 1978), 79–101.
5. Howard Dorgan, *The Old Regular Baptists of Central Appalachia: Brothers and Sisters in Hope* (Knoxville: Univ. of Tennessee Press, 1989), 21.
6. Johnny Ward, interview with author, Triplett, N.C., Jan. 20, 1991.
7. Howard Dorgan, *Giving Glory to God in Appalachia: Worship Practices of Six Baptist Subdenominations* (Knoxville: Univ. of Tennessee Press, 1987), 82–83. See also Jeff Todd Titon, *Powerhouse for God: Speech, Chant, and Song in an Appalachian Baptist Church* (Austin: Univ. of Texas Press, 1988), 311–16.
8. Dave Berkman, "Long Before Falwell: Early Radio and Religion—As

Reported by the Nation's Periodical Press," *Journal of Popular Culture* 21 (Spring 1988): 6; Quentin J. Schultze, "The Wireless Gospel: The Story of Evangelical Radio Puts Televangelism into Perspective," *Christianity Today* 32 (Jan. 15, 1988) 21.

9. Dean Fields, interview with author, Thornton, Ky., June 17, 1990; Brenda Blankenship, interview with author, Iaeger, W.Va., July 31, 1991.

10. Recorded Apr. 16, 1989.

11. *Giving Glory to God in Appalachia,* 173.

12. Roscoe Greene, interview with author, Deep Gap, N.C., Aug. 31, 1983.

13. Recorded Apr. 2, 1989.

14. This episode was observed March 12, 1989.

15. *Giving Glory to God in Appalachia,* 98–99.

16. Recorded Apr. 16, 1989, and May 26, 1991.

17. Recorded Aug. 4, 1991.

18. Garrett Mullins, interview with author, Grundy, Va., Apr. 2, 1989.

19. Dollie Shirley, interview with author, Boone, N.C., Aug. 4, 1991.

20. James H. Kelly, interview with author, Neon, Ky., May 21, 1989.

21. Recorded Apr. 2, 1989.

22. William Martin, "Perspectives on the Electronic Church," in *Varieties of Southern Religious Experience,* ed. Hill, 182–84; Quentin J. Schultze, "The Wireless Gospel," *Christianity Today* 32 (Jan. 15, 1988): 20–21.

23. Ramona Coles, interview with author, Beckley, W.Va., Aug. 25, 1990.

24. Roscoe Greene, interview with author, Deep Gap, N.C., Aug. 31, 1983.

25. Jane Smith, telephone interview with author, Aug. 28, 1983. Smith is the station manager at WATA.

26. See William Clements, "The American Folk Church," (Ph.D. diss., Indiana Univ., 1974); Bruce Rosenberg, *The Art of the American Folk Preacher* (New York: Oxford Univ. Press, 1970); Don Yoder, "Toward a Definition of Folk Religion," *Western Folklore* 33 (1974): 2–16; and Yoder, "Official Religion vs. Folk Religion," *Pennsylvania Folklife* 15 (1966): 36–52.

27. Elaine J. Lawless, *God's Peculiar People: Women's Voices and Folk Tradition in a Pentecostal Church* (Lexington: Univ. Press of Kentucky, 1988), 1–9.

28. Harrell, 27.

29. "Unusual, But Not Crazy," *Newsweek,* Dec. 19, 1988, 47.

30. "Broadcasters Comment on Strategies to Save AM," *Broadcasting* 115 (Aug. 29, 1988): 54–55; "Can AM Radio Be Saved?" *Broadcasting* 117 (July 3, 1989): 20–21.

31. *Newsweek,* Dec. 19, 1988, 47.

32. "Program Syndication: Pushing Radio's Hot Buttons," *Broadcasting* 119 (July 23, 1990): 35–55.

33. "Much Ahead for AM, FM," *Broadcasting* 119 (Oct. 1, 1990): 45; "Digital Audio Broadcasting: Choosing a Terrestrial or Satellite System?" *Broadcasting* 119 (Dec. 31, 1990): 60–62.
34. "Turn Your Radio On," in *Giving Glory to God in Appalachia*, 170–84.
35. Recorded Aug. 11, 1991.
36. Cleo Tester, telephone interview with author, Aug. 13, 1991.
37. Recorded Aug. 11, 1991.
38. Jim Jernigan, interview with author, Boone, N.C., July 25, 1991.
39. Fran Atkinson, interview with author, Mountain City, Tenn., Jan. 21, 1991.
40. Sam Sidote, interview with author, Welch, W.Va., Feb. 12, 1989.
41. From promotional materials currently distributed by WAEY.

2. Brother Johnny Ward and "The Voice of the Word"

1. My first visit was on November 18, 1973.
2. Dorgan, *Giving Glory to God in Appalachia*, xi, 55.
3. Mary Lou Hayworth, interview with author, Mountain City, Tenn., July 22, 1990.
4. Fran Atkinson, interview with author, WMCT, Mountain City, Tenn., Jan. 21, 1991.
5. Dorgan, *Giving Glory to God in Appalachia*, 175–76.
6. Lawless, *God's Peculiar People*, xiv–xv. Throughout this work, Lawless has much to say about the place of women in the Pentecostal tradition.
7. The other broadcast dates were November 18, 1973; July 17, 1983; July 22, 1990; July 29, 1990; and January 20, 1991.
8. Dorgan, *Giving Glory to God in Appalachia*, 40, 84.
9. Johnny Ward, interview with author, Triplett, N.C., Jan. 4, 1991.
10. Fran Atkinson, interview with author, Mountain City, Tenn., Jan. 21, 1991.
11. Johnny Ward, interview with author, Triplett, N.C., Jan. 4, 1991.
12. Fran Atkinson, interview with author, Mountain City, Tenn., Jan. 21, 1991.
13. I am indebted to Sadie Ward for the use of eleven tapes of broadcasts I did not witness: Mar. 22 and 29, 1990; June 24, 1990; Sept. 9, 16, and 30, 1990; Aug. 12, 19, and 26, 1990; and Oct. 7 and 14, 1990.
14. Recorded July 22, 1990.
15. Ibid.
16. Ibid.
17. "Full gospel" is a Pentecostal term. For further explanation see Donald W. Dayton, *Theological Roots of Pentecostalism* (Metuchen, N.J.: Scarecrow Press, 1987), 18. For an understanding of the "second-blessing"

theology see Lawless, *God's Peculiar People,* 26. The "Jesus-only" or "Oneness" movement in Holiness-Pentecostal history is detailed by Robert Mapes Anderson, *Vision of the Disinherited: The Making of American Pentecostalism* (New York: Oxford Univ. Press, 1979), 176–88.

18. These quotations and much of the information that follows came from an interview conducted at the Ward home in Triplett, N.C., on January 4, 1991. All six of the adults featured in this discussion were present for that interview.

19. Johnny and Sadie Ward, letter to author, Jan. 21, 1991.

20. "Randy Smith Revivals Newsletter" (n.p., n.d.).

21. Sadie Ward, "Voice of the Word Ministry: Jesus on the Inside . . . Working on the Outside," a tract (n.p., n.d.).

22. Troy D. Abell, "The Holiness-Pentecostal Experience in Southern Appalachia," in *Religion in Appalachia,* ed. Photiadis, 82.

23. Dorgan, *Giving Glory to God in Appalachia,* 56–77; and Dorgan, *The Old Regular Baptists of Central Appalachia,* 54–56.

24. In her study of southern Indian Pentecostals, Lawless did find chanted testimonials and preaching, "complete with the gasp at the end of the line": Lawless, *God's Peculiar People,* 100.

25. Anderson provides an excellent examination of "tongues": Anderson, *Vision of the Disinherited,* 15–27. "Dancing in the Spirit" and "swooning in the Spirit" are mentioned by Lawless, *God's Peculiar People,* 52–53 and 105. In central Appalachia "swooning in the Spirit" is frequently known as being "slain in the Spirit."

26. An interesting account of one Holiness testimony has been provided by Ruel W. Tyson, Jr., in "The Testimony of Sister Annie Mae," in *Diversities of Gifts: Field Studies in Southern Religion,* ed. Ruel W. Tyson, Jr., James L. Peacock, and Daniel W. Patterson (Urbana: Univ. of Illinois Press, 1988), 105–25.

27. Dewey Ward, interview with author, Drexel, N.C., Jan. 4, 1991.

28. Dorgan, *The Old Regular Baptists of Central Appalachia,* 178–81.

29. Dallas Ramsey, interview with author, Ashcamp, Ky., Aug. 17, 1986.

30. Anderson, *Vision of the Disinherited,* 113.

31. David Edwin Harrell, Jr., "The Evolution of Plain-Folk Religion in the South, 1835–1920," in *Varieties of Southern Religious Experience,* ed. Hill, 41; Anderson, *Vision of the Disinherited,* 116–17.

32. Lawless, *God's Peculiar People,* 31.

33. Dayton, *Theological Roots of Pentecostalism,* 26–28.

3. Rex and Eleanor Parker and "Songs of Salvation"

1. Rex and Eleanor Parker, *Your Favorite Hymns* (n.p., n.d.).

2. Rex and Eleanor Parker, interview with author, Lerona, W.Va., Aug. 25, 1990. Unless otherwise identified, all quotations of Rex or Eleanor Par-

ker were recorded during this lengthy Saturday afternoon interview conducted at the Parkers' home in Lerona.

3. From the tape of the Apr. 16, 1989, broadcast of "Songs of Salvation."

4. Provided to the author by the management of WAEY. These program descriptions and all statistical information about the stations come from this packet of materials.

5. "He'll Set You Free," *Rex and Eleanor No. 1 Personal Song Book* (Princeton, W.Va.: 1946), 5. I am indebted to Ivan M. Tribe for this material.

6. See Ivan M. Tribe, "Rex and Eleanor Parker: The West Virginia Sweethearts," *Bluegrass Unlimited* 10 (Apr. 1976): 18–25; Tribe, "West Virginia Country Music During the Golden Age of Radio," *Goldenseal* 3 (July–Aug. 1977): 15–26; Tribe, *Mountaineer Jamboree: Country Music in West Virginia* (Lexington: Univ. Press of Kentucky, 1984), 98–103, and elsewhere; and Tribe and John W. Morris, *Molly O'Day, Lynn Davis, and the Cumberland Mountain Folks: A Bio-discography* (Los Angeles: John Edwards Memorial Foundation, Inc., 1975), 3–5.

7. Eleanor Parker, interview with author, Princeton, W.Va., Sept. 16, 1990.

8. Tribe, "Rex and Eleanor Parker: The West Virginia Sweethearts," 19; and Parker, interview with author, Lerona, W.Va., Aug. 25, 1990.

9. Recorded Apr. 16, 1989.

10. Recorded Aug. 25, 1990.

11. From Apr. 16, 1989, "Songs of Salvation" broadcast, WAEY; and from Aug. 25, 1990, "Songs of Salvation" broadcast, WJLS.

12. Recorded Aug. 5, 1990.

13. Eleanor Parker, interview with author, Princeton, W.Va., Sept. 16, 1990.

14. From Aug. 25, 1990, "Songs of Salvation" Broadcast, WJLS. Also see Tribe, "Rex and Eleanor Parker: The West Virginia Sweethearts," 19–20; Tribe, *Mountaineer Jamboree,* 126; and Tribe and Morris, *Molly O'Day, Lynn Davis, and the Cumberland Mountain Folks,* 4.

15. Tribe, "West Virginia Country Music During the Golden Age of Radio," 15–16.

16. Garland Hess, "Radio With a Capital R," *Goldenseal* 10 (Fall 1984): 62.

17. Tribe, "West Virginia Country Music During the Golden Age of Radio," 21.

18. Garret Mathews, "Country Radio: The Early Days of WHIS, Bluefield," *Goldenseal* 10 (Fall 1984): 60.

19. Lewis C. Tierney, "Mercer County—A New Section of the State," *The West Virginia Review* 21 (Aug. 1944): 8–9.

20. Eleanor Parker, interview with author, Princeton, W.Va., Sept. 16, 1990.

21. Rex and Eleanor Parker, *Our Favorite Hymns,* 1955 Edition (n.p., n.d.), 4.

22. Tribe and Morris, *Molly O'Day, Lynn Davis, and the Cumberland Mountain Folks,* 5–8.

23. Tribe, "Rex and Eleanor Parker: The West Virginia Sweethearts," 22.

Tribe has changed his mind about the date of these Coral label recordings. He now states that these records were cut January 7, 1952; letter from Tribe to author, Aug. 26, 1991.

24. Letter from Arnett and Ellen Hamrick to the Parkers, n.d.

25. Eleanor Parker, interview with author, Princeton, W.Va., Sept. 16, 1990.

26. *Bluefield Daily Telegraph*, Mar. 18, 1956.

27. From three undated clippings from the *Bluefield Daily Telegraph*, identified as published in 1947, 1949, and 1956, provided by Ivan M. Tribe.

28. From a December 1959 article in the *Bluefield Daily Telegraph*, an undated clipping provided by Ivan M. Tribe.

29. Quoted from the Parkers' performance at the Lord's Full Gospel Mission, Pilot, Virginia, Aug. 19, 1990.

30. From a December 1959 article from the *Bluefield Daily Telegraph*, an undated clipping provided by Ivan M. Tribe.

31. Rex and Eleanor Parker, *Your Favorite Hymns* (n.p., n.d.).

32. Ibid.

33. Tribe and Morris, *Molly O'Day, Lynn Davis, and the Cumberland Mountain Folks*, 10. The Church of God (Cleveland, Tennessee) is the second largest Pentecostal denomination in the South. It is distinct from and not to be confused with the Holiness denomination of the same name, the Church of God (Anderson, Indiana). In this volume, all mentions of the Church of God refer to the Cleveland, Tennessee, denomination.

34. Rex and Eleanor Parker, interview with author, Lerona, W.Va., Aug. 25, 1990.

35. Observations made during the following studio visits: to WAEY on Apr. 16, 1989; July 22, 1990; Aug. 5, 1990; and Sept. 16, 1990; to WJLS on Aug. 25, 1990.

36. Unless otherwise noted, the "Songs of Salvation" quotations provided here have been taken from a tape of the April 16, 1989, broadcast.

37. Eleanor Parker, interview with author, Princeton, W.Va., Sept. 16, 1990.

38. David Price, interview with author, Princeton, W.Va., Sept. 16, 1990.

39. *Old Time Pentecostal Revival Songs* (Dallas Turner Publications, n.d.), Revivaltone Cassette 101, Side 1.

4. Brother Dean Fields, WNKY, and "The Words of Love Broadcast"

1. For more about "Bad John" Wright see Phillip K. Epling, *Bad John Wright: The Law of Pine Mountain* (Parsons, W.Va.: McClain Printing Company, 1981).

2. Ronald D. Eller, *Miners, Millhands, and Mountaineers* (Knoxville: Univ. of Tennessee Press, 1982), 144–45; Alphonse F. Brosky, "Building a Town for a Glimpse of Jenkins," *Coal Age* 23 (Apr. 5, 1923): 561–62.

3. Eller, *Miners, Millhands, and Mountaineers,* 144.
4. Robert M. Rennick, *Kentucky Place Names* (Lexington: Univ. Press of Kentucky, 1984), 211.
5. Ibid.
6. Unidentified convenience store owner, interview with author, McRoberts, Ky., May 26, 1991.
7. U.S. Census, 1940, 1950, 1960, and 1970.
8. Unidentified convenience store owner, interview with author, McRoberts, Ky., May 26, 1991.
9. U.S. Census 1980 and 1990.
10. I. A. Bowles, *History of Letcher County, Kentucky: Its Political and Economic Growth and Development* (Lexington, Ky.: Hurst Printing Company, 1949), 6–10, 16–17, 26–27, 41–42, 44, 66–67, 72.
11. All observations relative to WNKY were drawn from station visits on May 21, 1989, and May 26, 1991.
12. Wiley Vanover, interview with author, Neon, Ky., May 21, 1989.
13. Teddy Kiser, interview with author, Neon, Ky., May 26, 1991.
14. Interview, May 26, 1991. Name of interviewee not recorded.
15. Teddy Kiser, interview with author, Neon, Ky., May 21, 1989.
16. Epling, *Bad John Wright,* 6.
17. Howard Dorgan, *Giving Glory to God in Appalachia,* 56–66.
18. Taped May 21, 1989. In author's private collection.
19. James A. Kelly, interview with author, Neon, Ky., May 21, 1989.
20. All biographical information on Dean Fields has been drawn from two interviews, June 16, 1990 and May 25, 1991.
21. Rennick, *Kentucky Place Names,* 267.
22. Geraldine Boggs, interview with author, Thornton, Ky., May 25, 1991.
23. Martha Adams, interview with author, Thornton, Ky., May 25, 1991.
24. Brother James Simms, quoted in Dorgan, *Giving Glory to God in Appalachia,* 78–79.
25. Dean Fields, interview with author, Thornton, Ky., May 25, 1991.
26. See John B. Boles, *The Great Revival, 1787–1805* (Lexington: Univ. Press of Kentucky, 1972), 67–68; Catharine C. Cleveland, *The Great Revival in the West, 1797–1805* (Chicago: Univ. of Chicago, 1916; Gloucester, Mass.: Peter Smith, 1959), 57–60; Fredrick Morgan Davenport, *Primitive Traits in Religious Revivals* (New York: Macmillan Company, 1917), 73–86; William Warren Sweet, *Religion on the American Frontier: The Baptists, 1783–1830* (New York: Cooper Square Publishers, 1964), 615–16.
27. All quotations in this sequence are taken from May 25, 1991, visitation notes.
28. Dean Fields, interview with author, Thornton, Ky., May 26, 1991.
29. Dean Fields's sermon, June 17, 1990.
30. Dean Fields, interview with author, Thornton, Ky., May 25, 1991.

31. Dean Fields's sermon, June 17, 1990.
32. Refer to scene described by the author in *The Old Regular Baptists of Central Appalachia,* 22–23.
33. Taped May 26, 1991. In author's private collection.
34. Elder Earl Sexton, Silas Creek Union Baptist Church, Lansing, N.C., July 24, 1984.

5. Sister Brenda Blankenship, WELC, and Women in the Airwaves of Zion

1. The following sources disagree concerning the year of incorporation: Quinith Janssen and William Fernbach, *West Virginia Place Names* (Shepherdstown, W.Va.: J and F Enterprises, 1984), 83; *McDowell County History,* compiled by the Colonel Andrew Donnally Chapter of the Daughters of the American Revolution (Fort Worth, Tex.: University Supply and Equipment Company, 1959), 24–25. A public relations package supplied by the town of Welch gives July 12, 1894, as the incorporation date, and I assume that document to be correct.
2. Janssen and Fernbach, 83.
3. Ronald D. Eller, *Miners, Millhands, and Mountaineers: Industrialization of the Appalachian South, 1880–1930* (Knoxville: Univ. of Tennessee Press, 1982), 73–75.
4. Ibid., 73–74.
5. Phil Conley, *History of the West Virginia Coal Industry* (Charleston, W.Va.: Education Foundation, Inc., 1960), 236–37.
6. *West Virginia: A Guide to the Mountain State,* compiled by the Writers' Project of the Works Progress Administration (New York: Oxford Univ. Press, 1941), 476.
7. *McDowell County History,* 29.
8. At the time of this writing only preliminary 1990 United States Census counts had been released. Therefore, the 3,028 figure could be revised at a later date.
9. Mimeographed public relations packet distributed by the town of Welch.
10. At the time of this writing, raw 1990 population distributions had been released, but other census data were still unavailable.
11. U.S. Census statistics, 1960–90.
12. Conley, *History of the West Virginia Coal Industry,* 237.
13. Otis K. Rice, *West Virginia: A History* (Lexington: Univ. Press of Kentucky, 1985), 192.
14. Ibid., 193.
15. Mimeographed public relations packet distributed by the town of Welch.
16. Rice, *West Virginia,* 193.

17. Mimeographed public relations packet distributed by the town of Welch.
18. Martha Moore, interview with author, Welch, W.Va., Aug. 1, 1991.
19. Margaret Henrichs, interview with author, Welch, W.Va., Aug. 1, 1991.
20. Martha Moore, interview with author, Welch, W.Va., Aug. 1, 1991.
21. Sam Sidote, interview with author, Welch, W.Va., Aug. 1, 1991.
22. Sam Sidote, interview with author, Welch, W.Va., Feb. 12, 1989.
23. Brenda Blankenship, interview with author, Welch, W.Va., Feb. 12, 1989.
24. Ibid.
25. Kathy Benfield sermon, Drexel, N.C., Jan. 4, 1991.
26. Brenda Blankenship, telephone interview with author, Aug. 23, 1991.
27. Ibid.
28. Ibid.
29. Recorded Feb. 12, 1989, Welch, W.Va..
30. Recorded Sept. 9, 1990, Welch, W.Va..
31. Recorded Sept. 18, 1990, Kernersville, N.C..
32. Brenda Blankenship, telephone interview with author, Aug. 12, 1991.
33. Ibid.
34. Ibid.
35. Timmy Nipper, interview with author, Iaeger, W.Va., July 31, 1991.
36. Ibid.
37. Brenda Blankenship, telephone interview with author, Aug. 12, 1991.
38. Ibid.
39. Roscoe Greene, interview with author, Deep Gap, N.C., Aug. 31, 1983.
40. Ralph Smith, interview with author, Welch, W.Va., Aug. 1, 1991.
41. Rex and Eleanor Parker, interview with author, Lerona, W.Va., Aug. 25, 1990.
42. William L. Kell, Jr., interview with author, Welch, W.Va., July 31, 1991.
43. Ralph Smith, interview with author, Welch, W.Va., Aug. 1, 1991.

6. Some Common Threads

1. Loyal Jones, "Mountain Religion: The Outsider's View," *Religion in Appalachia,* ed. Photiadis (Morgantown: West Virginia Univ. Press, 1978), 401.
2. Ibid.
3. Howard Dorgan, "'He Will Give You a Sermon': Impromptu Preaching in the Old Regular Baptist Church," in *Religion,* vol. 5 of *Cultural Perspectives on the American South,* ed. Charles Reagan Wilson (New York: Gordon and Breach, 1991), 71–73.
4. Eleanor Parker, interview with author, Princeton, W.Va., Sept. 16, 1990.
5. Brenda Blankenship, telephone interview with author, Aug. 12, 1991.
6. Loyal Jones, "Mountain Religion," 406.

7. Eleanor Parker, "He'll Set You Free," *Rex and Eleanor No. 1 Personal Song Book* (Bluefield, W.Va.: 1946),5.

8. Rex and Eleanor Parker, interview with author, Lerona, W.Va., Aug. 25, 1990.

9. Ward family, interview with author, Triplett, N.C., Jan. 4, 1991.

10. Ibid.

11. Nathan L. Gerrard, "Churches of the Stationary Poor in Southern Appalachia," in *Religion in Appalachia,* ed. Photiadis, 274.

Index

The Airwaves of Zion was composed by The Composing Room of Michigan, Inc., printed by Braun-Brumfield, Inc., and designed by Sheila Hart. Text is set in New Century Schoolbook and display lines are Americana. The book is printed on 60# Glatfelter B-16 Natural Smooth.